Texturising

Defects, Causes, Effects, Remedies and Prevention through Quality Management

Texturising

Defects, Causes, Effects, Remedies and Prevention through Quality Management

By
H. V. Sreenivasamurthy
and
B. Purushothama

CRC Press
Taylor & Francis Group
Boca Raton London New York

CRC Press is an imprint of the
Taylor & Francis Group, an **informa** business

WOODHEAD PUBLISHING INDIA PVT LTD

New Delhi

First published 2025
by CRC Press
4 Park Square, Milton Park, Abingdon, Oxon, OX14 4RN

and by CRC Press
2385 NW Executive Center Drive, Suite 320, Boca Raton FL 33431

CRC Press is an imprint of Informa UK Limited

British Library Cataloguing-in-Publication Data
A catalogue record for this book is available from the British Library

Print edition not for sale in South Asia (India, Sri Lanka, Nepal, Bangladesh, Pakistan or Bhutan).

ISBN13: 9781032955001 (hbk)
ISBN13: 9781032955018 (pbk)
ISBN13: 9781003585206 (ebk)

DOI: 10.1201/9781003585206

Typeset in Times New Roman
by Bhumi Graphics, New Delhi

WPI

Contents

Preface

The initial investigation on 'Defects in Texturising' started sometime in May 1998, when Mr A. R. Kulkarni, then Senior Vice President, Microsynth Fabric Ltd, Silvasa, evinced keen interest in the industry-oriented research project as above proposed by the author. The study was further extended in detail at Reliance Industries Ltd, Silvasa, till 2002 July. We gratefully acknowledge the initial support and co-operation extended by Mr Kulkarni to work on shop floor of Microsynth Fabric Ltd, Silvasa, and by Mr C. Bose, President, Reliance Industries Ltd, Silvasa, for providing all facilities in completing the research project.

We wish to record the co-operation and help extended by all the staff members of both the industries and acknowledge the sincere efforts put in by Mr Rahul, R. Dabhade and Mr A. K. Sinha in conducting experiment trials of the project during their undergraduate and postgraduate studies, respectively.

This book has been designed to serve the shop-floor personnel directly dealing with day-to-day production and quality problems and to device methods for preventing the defects in the texturised yarns. This book also explains various precautionary measures to be taken not only at manufacturing but also during handling, storing, packing, delivering and guiding the customers.

As the technology is improving, the defects relating to technical reasons are reducing, but the defects due to poor work practices are still there. This aspect is also discussed in this book.

The authors will be happy to receive criticism and suggestions for further improvement of this book.

The authors are very much thankful to Sri K. L. Vidur, who reviewed the complete book, suggested some changes and also wrote the Foreword.

H. V. Sreenivasamurthy
B 1 – 902, Elita Promenade
Opp. to RBI Layout, J P Nagar – 7th Phase
Bangalore – 56007, India

B. Purushothama
No 15. Sumithra
3rd Temple Street
15th Cross, Malleswaram
Bangalore – 560003, India

Foreword

The man-made continuous filament industry was started by the nitrocellulose, viscose, cuproammonium and actate of rayon family of continuous filaments followed by the invention of the polyamide nylon 66 by Dr W. H. Carothers of E. I. du Pont de Nemours, USA, in 1933 and commercialised to produce nylon 66 continuous filament yarn in the early 1940s. Since then, the synthetic fibre industry has grown into various types of chemical fibres such as polyamide, polyester and acrylic for a multifunctional application and end use. The basic continuous filament yarns were lacking some of the functional properties of the yarns of natural fibres such as cotton, wool and silk for the apparel and other technical applications. Hence, this necessitated the technologists to invent/develop the process for the yarn modification using the thermoplastic properties of the synthetic filament yarn to a texturised/crimped yarn.

The book is a blend of an academician and an industry personnel with their vast knowledge of the subject in the field and experience of association with the industry. The authors have covered all the aspects starting from the origin of the texturising technology of yarn modification, type of machinery, processs, process control, testing and quality control and the Management Information Systems. The authors have covered the most important aspect, the problems and remedies in the texturising industry.

The book is a good guide for the students studying man-made fibres/ textiles in which texturising is an important subject.

As most of the units of texturising in India are based on the raw materials POY and FDY supplied by the filament yarn industry, this book is an useful and informative guide for the day-to-day problems faced by the shop-floor personnel.

However, the problems, defects and remedies faced in the texturising industry vary from unit to unit, type of machine and raw material; this book is a result-oriented guide to the shop-floor personnel in the Indian texturising industry.

With best wishes.

K.L. Vidurashwatha (Vidur)
M. Tech, FIE, Hon. FTA-2012
Chartered Engineer

Introduction

Synthetic filaments were developed mainly to replace costly and scare natural fibres. They have the advantage of uniformity, possibility of engineering the fibre by controlling the denier and bulk, developing the tenacity and elasticity by modifying the structure and so on. Different synthetic filaments were developed in last one century. Rayon was the first man-made filament fibre, developed in the 1890s. Nylon was first produced in the late 1930s and soon followed by the polyesters and polyacrylics. Polyolefins and polyurethanes were produced during the 1950s. Today, almost all the manufactured fibres can be used in filament form in filament yarns. Acrylic and modacrylic are exceptions – they are typically used in staple form. Filament yarns are composed of one or many continuous filaments. Filament yarns are assembled with or without twist.

Polyester occupies a distinct position among the textiles in terms of its versatility and widespread usage in apparel and industrial applications. But flat filament polyester is uncomfortable for apparel use as it lacks the comfort properties like that of staple yarn from natural origin. So to introduce the comfort and aesthetics in flat filament synthetic yarn, 'texturising' is the most widely used technique in the industry.

The yarns available can be grouped as follows:

- Flat continuous filament yarns can be monofilament, which contain only one single continuous filament, or multifilament yarns with a number of filaments of fine deniers combined to form one continuous strand of filaments varying in range from fine to coarse counts.
- Flat multifilament yarns create a very smooth fabric surface. Round cross-sectional shaped fibres are typically used in filament yarns, producing very compact and smooth yarns. End uses for these yarns include seat belts, tire cord and tire cord fabrics. Apparel fabrics needing closely packed yarns for reduced air and water vapor permeability also use flat multifilament yarns. This type of filament yarn has elongation, depending on the fibre type that is used.
- Monofilament yarns are not normally used for apparels as they are not comfortable, but there are some exceptions of end uses as a dress material. They are used mainly for technical purposes. Fishing line is a good example of a monofilament yarn.

- Texturised continuous filament yarns have the filaments modified in some way to give bulk and/or specific stretch properties. Texturised filament yarns have a noticeably greater volume than conventional filament yarns of similar filament count and linear density. The yarns have a relatively low elastic stretch, but can be developed specifically to impart stretch. The yarn is sufficiently stable to withstand wet finishing. The apparent volume of the yarn is achieved through physical, chemical or heat treatments, or a combination of these.
- Spun yarns are formed using short or non-continuous fibres. The yarn is held together by twisting the fibres to produce necessary fibre-to-fibre cohesion and adequate yarn strength.
- Short staple spun yarns utilise fibres less than 64 mm in staple length. Normally, yarns spun from short staple are more hairy, irregular and need higher twist.
- Long staple spun yarns utilise fibres from 64 mm (2.5 inches) up to 460 mm (18 inches) in length. However, the most common range of staple length is 64 mm (2.5 inches) to 152 mm (6 inches).

Filament yarns can be classified as flat or texturised. Flat filament yarns have all their filaments aligned parallel to the yarn axis. The texturised yarns have individual filaments lying in a curled, twisted, snarled, looped or crimped configuration. The most common methods of texturising are false-twist, air jet and stuffer box. These methods will be discussed later in the module.

Texturising techniques were initially relevant to artificial fibres to diminish characteristics such as glassiness, slipperiness and the chance of pilling (configuration of little fibre tangles on a fabric shell). They make yarns more opaque while improving look and touch and boost warmness and absorbency. It is by the formation of folds, loops, coils or crinkles in filaments. Such alterations in the physical shape of a fibre have an effect on the behaviour and hand of textiles made from them.

Among different types of texturising techniques, 'false-twist texturising' is the most popular in texturising industry. It is basically a thermomechanical deformation process which changes the characteristics of filament yarn to generate bulk and warmth. Another popular method is air texturising, which uses air jet to get entanglement and also bulk to impart spun yarn effect.

During texturising process, incidentally lots of defects occur. The help of students were taken to study different functional defects in false-twist texturised yarns generated in actual mill working conditions. A systematic and analytical approach was followed to track the causes of the defects. The causes ranged from feed material characteristics, various mechanical defects, machine parts, process parameters and so on. At the end, practically achievable remedial measures were also suggested to obtain fault-free consistent end product. The findings were further validated by referring to various researches and are presented as a part of this book in Chapter 4.

The term 'texture' is derived from the Latin word 'Texere' and it describes those attributes of an object that can be recognised by the human site (visual characteristics) or touch (tactile characteristics). In textile fabrics, warmth, opacity, flexibility, good wearing properties coupled with a characteristic appearance and easy maintenance are the main texture properties desired.

All these desirable properties are inherently provided with yarn spun from staple fibres, whereas flat synthetic filament yarns lack the human appreciation though they possess some superior properties, namely, high strength and good elasticity. The primary object of a flat filament yarn conversion process is to initiate the desirable properties of normal spun yarn. One method of achieving this is to cut the continuous filaments into staple and then process them into yarn using conventional or non-conventional spinning systems. But this process is time-consuming and involves a number of processes which are costly. On the other hand, continuous filaments can be converted into yarns by different methods at lower cost in a single process called 'texturising'.

The texturised yarns are those yarns that have been processed to introduce double crimps, coils, loops or any other fine distortions along the length of the filament. Thus, by texturising, the filament yarns attain soft and 'natural like' feel and increased warmth and comfort properties.

The fabrics containing texturised yarns have increased bulk, opacity, moisture absorbency and improved thermal insulation properties with warmer handle. Some texturised yarns also confer extensible or 'stretch' properties on fabric made from them.

Texturising being a relatively new technology compared with traditional spinning and twisting of natural staple fibres, much literature is not available to educate the new comers on the possible defects, their causes, the effects and the suggested remedial actions. Hence, an attempt is made to compile the defects and their causes, so as to educate the students and new comers in the field.

In the foregoing pages, the defects observed in the most popular and most commercially exploited 'draw texturised yarn' are detailed with definition, causes, effects and remedies. Wherever possible the photographs of the defects are given.

Attending and correcting a defect is one aspect of working to satisfy the customers, whereas it is also needed to assure the quality by following suitable systems by committed employees. The next chapter, therefore, explains the system to be implemented to ensure defect-free yarns, which include the concepts of Quality Management System, Work Quality Management and Total Quality Management followed by a chapter on testing methods normally followed for the texturised yarns. The importance of educating the people on the spot on the customer requirements and also on the technical aspects following good practices of housekeeping maintenance and safety management is also discussed.

In the end, standard testing methods as specified by standard laboratories are given as appendixes.

1

Abbreviations and Explanations of Terms Used

Some of the abbreviations used in this book are explained below.

Abbreviation	Full Form
ATY	Air-texturised yarn
CV	Co-efficient of variation
DR	Draw ratio
DTY	Draw texturised yarn
DY	Ratio of disc surface speed to yarn speed in metre per minute
EMS	Environment Management System
FOY	Fully oriented yarn
IV	Intrinsic viscosity
LOY	Low oriented yarn
MSDS	Material Safety Data Sheet
OPU	Oil pick-up
Positorq	It is friction twisting unit provided in draw texturising machine
POY	Partially oriented yarn
QMS	Quality Management System
Roll-1	First feed unit (before primary heater)
Roll-2	Second feed unit (after friction unit and before setting zone)
Roll-3	Third feed unit (after setting zone before take-up)
T1	Input slide yarn tension before false-twist aggregate
T2	Output side yarn tension (immediately after friction unit)
T3	Yarn tension before take-up
T min	Minimum temperature
T opt	Optimum temperature
THUT	Twist–heat-set–untwist
TQM	Total Quality Management
UDY	Undrawn yarn (same as unoriented yarn)
UOY	Unoriented yarn
V/R	Velocity ratio
WQM	Work Quality Management

Explanation of some of the terms used in texturising industry

Some of the terms normally used in texturising are explained below.

- ***Abnormal crimp***: A relative term for crimp that is either too low or too high in frequency and/or amplitude or that has been put into the fibre with improper angular characteristics.
- ***Abraded yarns***: A filament yarn in which filaments have been cut or broken to create hairiness (fibrillation) to simulate the surface character of spun yarns. Abraded yarns are usually plied or twisted with other yarns before use.
- ***Air jet texturising***: A method of giving texturised effect by mechanical interlocking using strong air currents, not by heat-setting. In this method of texturising, the yarn is led through the turbulent region of an air jet at a rate faster than it is drawn off on the far side of the jet. In the jet, the yarn structure is opened, loops are formed and the structure is closed again. Some loops are locked inside and others are locked on the surface of the yarn. An example of this method is the Taslan process.
- ***Alternating twist***: A texturising procedure in which S- and Z-twists are alternately inserted in the yarn by means of a special false-twist spindle drive.
- ***Core spinning***: The process of making a core-spun yarn. It consists of feeding the core yarn (an elastomeric filament yarn, a regular filament yarn, a texturised yarn or a previously spun yarn) into the front delivery roll of the spinning frame and of covering the core yarn with a sheath of fibres during the spinning operation.
- ***Core-spun yarn***: A yarn made by twisting fibres around a filament or a previously spun yarn, thus concealing the core. Core yarns are used in sewing thread, blankets and socks and also to obtain novelty effects in fabrics.
- ***Crimp amplitude***: the height of displacement of the fibre from its un-crimped condition.
- ***Crimp frequency***: The crimp level, or the number of crimps per inch in yarn or tow.
- ***Crimp setting***: An after-treatment to set the crimp in yarn or fibre. Usually, heat and steam are used, although the treatment may be chemical in nature.
- ***Crimp***:
  1. The waviness of a fibre expressed as crimps per unit length. It is the difference in the distance between the two points on an un-stretched fibre and the same two points when the fibre is straightened under specified tension. Crimp is expressed as a percentage of the un-stretched length.

2. The difference in the distance between the two points on a yarn as it lies in a fabric and the same two points when the yarn has been removed from the fabric and straightened under specified tension, expressed as a percentage of the distance between the two points as the yarn lies in the fabric.

- ***Crimping***: The process of imparting crimp to tow or filament yarn.
- ***D/Y ratio***: The D/Y ratio is the ratio of speeds between the friction discs and the linear speed of the yarn. The most common method of changing the amount of twist inserted in the yarn is by changing the speed of the discs, keeping the number of discs and the spacing between them constant.
- ***Decitex***: A number indicating the weight in grams of 10,000 m of a specific yarn.
- ***Denier***: A number indicating the weight in grams of 9000 m of a specific yarn.
- ***Denier per filament (dpf)***: The denier of an individual continuous filament or an individual staple fibre. In filament yarns, it is the yarn denier divided by the number of filaments.
- ***Different texturised yarns***: Loopy yarn, crimped yarn, crindle yarn, stretch yarn, set yarn, alternate twist yarn, crinkled yarn.
- ***DR***: The DR is the amount of yarn stretched between the input roller and the output roller which passes through the primary heater and the twisting mechanism.
- ***Draw texturising***: A process in which the drawing stage of synthetic filament yarn manufacture is combined with the texturising process on one machine. Note: The drawing and texturising stages may take place in separate, usually consecutive, zones of a machine (sequential draw texturising) or together in the same zone (simultaneous draw texturising). In the manufacture of thermoplastic fibres, the simultaneous process of drawing and texturising is to increase molecular orientation and imparting crimp to increase bulk.
- ***Draw twisting***: The operation of stretching continuous filament yarn to align and order the molecular and crystalline structure in which the yarn is taken up by means of a ring-and-traveler device that inserts a small amount of twist (usually 8–10 turns/m) into the drawn yarn and wound on cops or tubes.
- ***Edge crimping method***: In this method of texturising, the thermoplastic yarns in a heated and stretched condition are drawn over a crimping edge and cooled. Edge-crimping machines are used to make Agilon yarns.
- ***False-twist***: The twist inserted in a yarn using false-twisting such that the net twist in the yarn is zero.

- ***False-twisting:*** A twisting operation applied at an intermediate position on a yarn so that no net twist can be inserted, as distinct from twisting at the end of a yarn where real twist is inserted.
- ***False-twist method of texturising***: A process in which a single-multifilament yarn is twisted, set and untwisted. When yarns made from thermoplastic materials are heat-set in a twisted condition, the deformation of the filaments is 'memorised' and the yarn is given greater bulk. This continuous method for producing the texturised yarns utilises simultaneous twisting, heat-setting and untwisting. The yarn is taken from the supply package and fed at controlled tension through the heating unit, through a false-twist spindle or over a friction surface that is typically a stack of rotating discs called an aggregate, through a set of take-up rolls, and onto a take-up package. The twist is set into the yarn by the action of the heater tube and subsequently is removed above the spindle or aggregate resulting in a group of filaments with the potential to form helical springs. Much higher processing speeds can be achieved with friction false-twisting than with conventional spindle false-twisting. Both stretch and bulked yarns can be produced by either process. Examples of false-twist texturised yarns are Superloft®, Flufflon® and Helanca®.
- ***FDY:*** Fully drawn yarn.
- ***Flat continuous filament yarns***: Flat continuous filament yarns are formed using very smooth and straight continuous filaments. Can be multifilament or monofilament.
- ***Gear crimping method***: In this texturising method, yarn is fed through the meshing teeth of two gears. The yarn takes on the shape of the gear teeth.
- ***Heat-setting***: A heat treatment given to thermoplastic materials to set the twist/shape and dimensional stability.
- ***High shrink staple:*** Staple with a higher degree of potential shrinkage than regular staple of the same generic fibre. When blended with regular staple and treated (in yarn or fabric form) to induce shrinkage, it produces a high degree of bulk in the product.
- ***Knit-de-knit method***: A type of yarn texturising in which a crimped yarn is made by knitting the yarn into a fabric, and then heat-setting the fabric. The yarn is then unraveled from the fabric and used in this permanently crinkled form. The yarn is knit into a 51-mm (2-inch) diameter hose-leg, heat-set in an autoclave and then unraveled and wound onto a final package.
- ***Long staple spun yarns***: Long staple spun yarns utilise fibres from 64 mm (2.5 inches) up to 457 mm (18 inches) in length in some cases.

- ***LOY***: Least oriented yarn. It is a continuous filament yarn spun at speeds below 1000 m/min.
- ***POY***: Filament yarns in which partial drawing is done (the DR is less than normal) resulting in partial longitudinal orientation of the polymer molecules.
- ***Set yarn***: A texturised yarn that is heat relaxed to reduce torque. False-twist yarns stabilised to produce bulk.
- ***Short staple spun yarns***: Short staple spun yarns utilise fibres less than 64 mm (2.5 inches) long.
- ***Skein***: A continuous strand of yarn or cord in the form of a collapsed coil. It may be of any specific length and is usually obtained by winding a definite number of turns on a reel under prescribed conditions. The circumference of the reel on which yarn is wound is usually 1 m.
- ***Spun yarns***: Spun yarns are formed using short or non-continuous fibres.
- ***Staple fibres***: Fibres in a tow are cut into fixed short length.
- ***Stretch yarn***: Qualitative term to describe a texturised yarn. A stretch yarn develops more stretch than bulk in the finished fabric.
- ***Stuffer box method***: The crimping unit consists of two feed rolls and a brass tube stuffer box. By compressing the yarn into the heated stuffer box, the individual filaments are caused to fold or bend at a sharp angle, while being simultaneously set by a heating device.
- ***Tex***: A number indicating the weight in grams of 1000 m of a specific yarn.
- ***Texture***: A term describing the surface effect of a fabric, such as dull, lustrous, wooly, stiff, soft, fine, coarse, open or closely woven; the structural quality of a fabric.
- ***Texturised continuous filament yarns***: Texturised continuous filament yarns are formed using continuous filaments that have been modified to give bulk and/or stretch properties.
- ***Texturised yarn:*** A continuous filament yarn that has been processed to introduce durable crimps, coils, loops or other fine distortions along the lengths of the filaments.

  Note 1: The main texturising procedures which are usually applied to continuous filament yarns made from or containing thermoplastic fibres, are as follows:

  i. The yarn is highly twisted, heat-set and untwisted either as a process of three separate stages (now obsolescent) or as a continuous process (false-twist texturising). In an infrequently used alternative method, two yarns are continuously folded together, heat-set and then separated by unfolding.

ii. The yarn is injected into a heated stuffer box either by feed rollers or through a plasticising jet of hot fluid (invariably air or steam). The jet process is sometimes known as jet texturising, hot air jet texturising or steam jet texturising.

iii. The yarn is plasticised by a passage through a jet of hot fluid and is impacted on to a cooling surface (impact texturising).

iv. The heated yarn is passed over a knife-edge (edge crimping) (now obsolete).

v. The heated yarn is passed between a pair of gear wheels or through some similar device (gear crimping).

vi. The yarn is knitted into a fabric that is heat-set and then unraveled (knit-de-knit texturising).

vii. The yarn is overfed through a turbulent air stream (air texturising, air jet texturising), so that entangled loops are formed in the filaments.

viii. The yarn is composed of the bicomponent fibres and is subjected to a hot and/or wet process whereby differential shrinkage occurs

Note 2: Procedures (i) and (iv) in Note 1 above give yarns of a generally high-stretch character. This is frequently reduced by re-heating the yarn in a state where it is only partly relaxed from the fully extended condition, thus producing a stabilised yarn with the bulkiness little reduced but with a much reduced retractive power.

Note 3: The procedure (vii) may also be applied to fibres which are not thermoplastic.

- ***Texturised***: An adjective used to describe continuous filament manufactured yarns (and woven and knit fabrics made therefrom) that have been crimped or have had random loops imparted, or that have been otherwise modified to create a different surface texture.
- ***Texturising***: The process of crimping, imparting random loops or otherwise modifying continuous filament yarn to increase cover, resilience, abrasion resistance, warmth, insulation and moisture absorption or to provide different surface texture that can reduce yarn lustre. A process performed on specialised machinery which creates bulk, stretch to the yarn, and therefore creates new aesthetics to the finished fabric.
- ***Torque yarn***: When a twisted yarn is permitted to hang freely, it rotates or kinks to relieve the torque introduced into the yarn during texturising.
- ***Tow***: Filaments from multiple spinnerets form a tow.
- ***Yarn lustre***: Referred to in terms such as bright, semi-dull, dull or full dull.

2
Brief introduction to Texturising

The man-made fibres were developed basically to replace naturals, which were becoming scarce and costly. However, the wide difference in the properties and appearance between the synthetic filaments and spun yarns with natural spun yarns prevented the former from competing effectively with natural spun yarns.

The production of texturised yarns stemmed from the need to broaden the area of use of synthetic fibres; applications of such fibres had previously been limited by the fibres because of low hygroscopicity and smooth surfaces, which had an unpleasant, glossy sheen. However, the texturising of the filament structure by simple mechanical or chemical distortion has enabled to stimulate spun yarn characteristics.

Texturising improves the use characteristics and hygienic properties of synthetic yarns. The texturised filament yarns were made suitable for knitting and weaving for use in apparels and clothing. In addition, some of their properties, in certain cases, are completely unique in the textile field and are creating a new 'customer appeal'. They are used for manufacturing a wide variety of textile products such as hosiery, knitted underwear, knitted outerwear shape retaining knitted fabrics, suits and overcoats, artificial fur, carpets, blankets and drapery and upholstery fabrics.

The term 'texture' defines and describes the attributes of an object that can be recognised by the human sight (visual characteristic) or touch (tactile characteristic). The main texture properties desired in apparel and domestic textiles are warmth, opacity, flexibility and good weaving/knitting properties, coupled with characteristic appearance and easy maintenance. These qualities are inherently provided with staple yarn spun from naturals, whereas synthetic filament yarns do not have all the properties desired.

The shiny lustrous synthetic filament plain fabric is not visually pleasing and disturbs the vision whereas the fabric made from cotton and wool is appealing to human eye. The superiority of these natural fibres is due to random hairiness of the spun yarns. These protruding hairs buckle under the effect of slight touch and are responsible for natural soft feeling. Air pockets are formed between these hairs which creates warmth of the fabric. The hairs also give bulkiness to the yarns and, in turn, increase the cover properties. The random hairs in natural spun yarns reflect light in every direction, which gets a diffused effect that appears matt and tend to be a natural part of the immediate environment. An array of closely packed filament provides a shiny surface because of the reflection of light by each filament which is parallel to the filament axis.

Texturising is the modification process of regular structure of synthetic filament to somewhat random structure by the formation of folds, loops, coils or crinkles. This is a process of crimping, imparting random loops or otherwise modifying continuous filament yarn to increase cover, resilience, abrasion resistance, warmth, insulation and moisture absorption or to provide a different surface texture. Such alterations in the physical shape have an effect on the behaviour and hand of textiles made from them.

The texturised yarns have bulk and stretch. Depending on the end use, the texturised yarns can be produced with low stretch and high bulk or vice versa. Sometimes a compromise is reached between the two properties and yarns can be made with moderate bulk. Thus, stretch and bulk are combined in the texturised yarn.

Texturising methods make yarns more opaque, improve look and touch and boost warmness and absorbency. The texturised yarns are synthetic uninterrupted filaments, modified to impart bulkiness, soft feel and crimp.

2.1 Types of Texturised Yarns

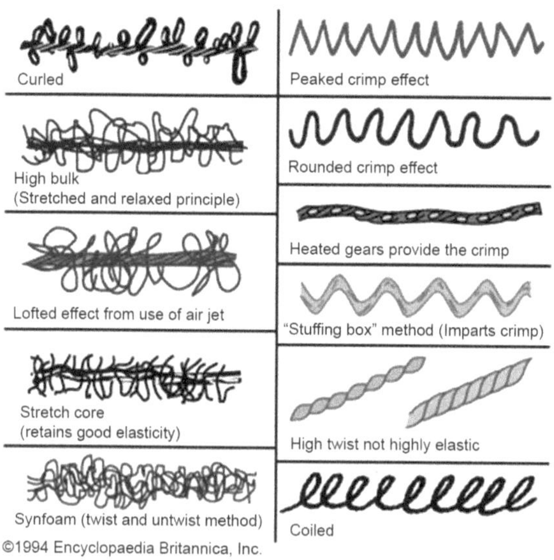

Fig. 2.1: Different texturised yarns

There are different varieties of the texturised yarns available and Figure 2.1 is just an illustration.

Various techniques are adapted for texturising synthetic filament yarns, basically utilising the thermoplastic properties of yarns either by the use of torque or by non-torque crimping devices. Texturising techniques have also been developed which do not use thermoplasticity of fibres. Different techniques used for texturising are given in Table 1.

The texturised yarns are basically of three types, and in each group there are variations. The basic types are the texturised yarns developed based on the thermoplastic nature of the filament, synthetic yarns are not developed based on the thermoplastic nature and the texturised yarns are natural and/or blends.

In texturising process making use of the thermoplastic properties of synthetics, the untexturised yarns are heated to plastic condition after they were distorted by texturising elements (spindle, gear, knife, etc.). Then they are cooled to retain the required shape. While texturising without using the thermoplastic properties, different techniques such as using air currents for texturising, core spinning with synthetics of different shrinking capacities and so on are popular.

Texturising techniques can also be broadly categorised as follows:

- Mechanical texturising
- Thermo-mechanical texturising
- Bicomponent texturising
- Differential shrinkage texturising
- Chemical texturising

There are two main methods of crimp configuration. They are as follows:

- Discontinuous method – conventional Helanca process (batch system) (THUT process).
- Continuous method
  - (a) False-twisting (THUT process method),
  - (b) Edge crimping method and
  - (c) Turbo-duo-twist method.

Let us have a brief discussion on various techniques used for texturising.

Table 1: Texturising Techniques

Synthetics			Natural and/or man-made
Based on the thermoplastic nature of the filament		**Not based on the thermoplastic nature of the filament**	
Torque crimping	**Non-torque crimping**		
1. Discontinuous method, i.e. Helanca multistep process 2. Continuous method a. False-twist texturising: rotor type and friction type b. Draw texturising: simultaneous and sequential c. Turbo-duo-twist method, i.e. twist texturising	1. Stuffer box method 2. Edge crimping method 3. Gear crimping method 4. Knit-de-knit method 5. Bicomponent fibres	1. Air texturising method 2. Combination texturising method (core spinning) 3. Composite yarn method (Bobtex) 4. Fibrillation technique	1. Slack mercerisation of yarns 2. Crimping and crosslinking 3. Thermoplasticisation of cellulose 4. Sheath–core technique 5. Crystallisation and de-crystallisation (solvent texturising) 6. Interfacial polymerisation 7. Differential shrinkage

2.1.1 Torque Crimping

2.1.1.1 Helanca Multistep Process

It is a discontinuous process adapted for high-tensile yarns. The twisting, heat-setting and untwisting operations are done in batches. Due to this discontinuous method, production was less. It is produced by the twisting of multifilament synthetics (polyamide, polyester and others) up to 2500–5000 twists/m, then heat-setting of the twisted yarn and untwisting of the set yarn. The yarn assumes a spiral shape and acquires high stretch ability (up to 400%) and fluffiness. High-tensile yarn is called *Elastik* in Russia and with a trade name of *Helanca* at other places.

In 1931, an American Rudolph H. Kägi developed a method of achieving the characteristics of wool in flat artificial silk fibres. He spun the acetate fibres into a springy spiral, and then wove it with wool to produce the first stretch fabric. The company Heberlein & Co. in Wattwil acquired this process and patented it. This new yarn, which no longer possessed the gloss and smoothness of synthetic silk, was soft and warm just like wool and was trademarked under the name 'Helanca'.

At first, the dimensional stability of the articles made from Helanca synthetic silk was inadequate. With each wash, an article of clothing became larger and longer, and elasticity was lost. In 1935, a new method of 'stabilising' the wool-like crimping with formaldehyde solved the problem of dimensional stability.

When World War II began, Europe was almost completely cut off from its textile raw material sources overseas. Therefore, artificial silk and rayon staple became important, and also the Helanca process. After World War II, Helanca initially lost its importance. During this time, the first deliveries of the new 'wonder material' nylon were arriving from the United States. In 1945, the first few kilograms of nylon yarn were available in Wattwil. Heberlein & Co. used the Helanca process to treat nylon, and the results were very satisfying. The new thread was 'fantastically elastic'. It was so elastic that knitters were against using this Helanca product.

In Lyon, France, the Helanca method was progressing in another direction; where before the nylon fibres were simply rolled up, fixated and uncoiled again, the French were processing two nylon threads – one coiled clockwise and another coiled counter-clockwise. Both threads were then twisted. The excessive stretching of nylon Helanca from a single thread was therefore remedied, and the elasticity of the new yarn reached optimal levels. This new texturised yarn caused a revolution in the hosiery market of the 1950s and 1960s. Hosiery produced with this new yarn could be dyed in more brilliant and glossy colours than with conventional yarns. Stockings and socks became more colourful, crease-free and easy to care for. A major advantage was that due to the high elasticity of this yarn, fewer stocking sizes had to be produced. This also simplified the production of tights which were beginning to appear. The first fine Lady's Helanca stockings on the market were crepe stockings, and a while later the first Helanca crepe tights appeared.

2.1.1.2 Continuous Method

False-Twist Method

False-twist texturising method is the most important and widely used technique for producing the texturised yarns. The operations of THUT are made simultaneously and continuous. The yarn is taken from the supply package and fed at controlled tension through the heating unit, through a false-twist spindle or over a friction surface that is typically a stack of rotating discs called an aggregate, through a set of take-up rolls, and onto a take-up package. The twist is set into the yarn by the action of the heater tube and subsequently is removed after the spindle or aggregate resulting in a group of filaments with the potential to form helical springs. Much higher processing speeds can be achieved with friction false-twisting than with conventional spindle false-twisting. Both stretch and bulked yarns can be produced by either process.

The yarns produced by the conventional THUT process generally have excellent characteristics; however, there are certain inherent drawbacks in the process itself.

(a) Slowness of the process because of the limitations imposed by the spindle speed of the up twister and the batch orientation of the process itself.

(b) The process requires extra care in handling the packages to avoid mixing of yarns.

(c) Uniformity is difficult to control between the packages and lots.

To overcome the above-mentioned objections, the continuous false-twist method is employed extensively in the production of high stretch and reduced stretch yarns. The false-twist method combines all the three stages, namely, twisting, heat-setting and untwisting, in one continuous operation. The yarn is drawn from the supply package, fed at controlled tension through tension guides, over the heater and through false-twist spindle, and finally wound on a package. The twist in the yarn is set when it is between the input feed roll and the false-twist spindle, by heating up and cooling before it leaves the false-twist spindle.

The mechanisms used for false-twisting are categorised as rotor-type mechanisms and friction-type mechanisms.

(a) ***Rotor-type false-twist mechanisms:*** The common feature of all rotor-type mechanisms for false-twisting is the use of a hollow spindle in the shape of a tube or rod of smaller diameter, which is called as false-twist spindle or revolving tube. Rotor-type false-twist mechanism works with ball bearings or with air cushion or rotated by discs. The speed of today's rotors have crossed 1000,000 RPM (1 million RPM).

The three basic steps of the twist–untwist method are carried out simultaneously and continuously as illustrated in the figure.

- Twist enters the filament yarn through the tensioner.
- Twist is set at the heater.
- The yarn is untwisted by the continuously rotating twist head.

The individual filaments snarl up protruding from the main body of the yarn. The yarn assumes the form of a twist-free bundle of twist lively filaments.

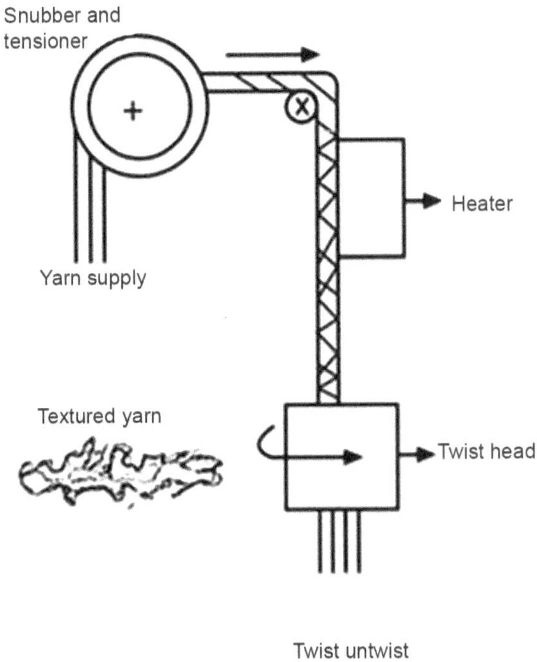

Fig. 2.2: Twist–untwist method of texturising

(b) ***Friction-type false-twist mechanisms:*** Friction-type mechanism finds an increasing use today because of very high speed. Endless belts, bushings, rings, discs and other rotating bodies may be used as twisting elements. The surfaces of all elements coming in contact with the yarn must possess a high friction factor. They are made of different materials such as diamond nickel-coated, solid ceramic and most frequently polyurethane to get a better quality at high speeds.

(c) ***Crossed belt twisting*** is another method of friction twisting. The yarn is held at the nip point of two belts mounted at an angle and moving in the opposite directions. These are referred to as nip twisters. Nip twisters were introduced in 1979. The rotation of the belts imparts both twist and a forwarding action to the yarn. The twist insertion rate is controlled by three factors: the angle of the belts to each other, the contact pressure applied to the belts and the V/R, which is the equivalent of the D/Y ratio for stacked discs. It is possible to run two yarns with opposite twist through a single unit and combine them at wind-up.

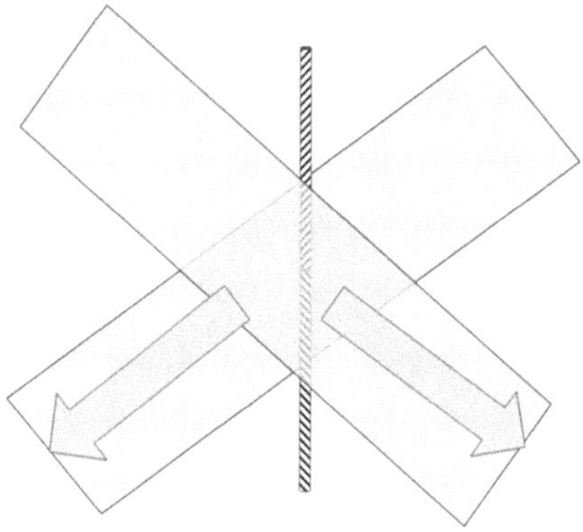

Fig. 2.3: Cross belt twisting

Draw Texturising (Friction Texturising)

Draw texturising is a process in which an undrawn flat yarn is drawn to the desired size and then texturised in a continuous manner. In the conventional method of texturising, that is, false-twist method, the feed yarn is fully drawn at spinners' end before it comes to texturising. The fully drawn yarn is obtained in two stages, namely, spinning and draw twisting or draw twisting which makes the process expensive and lengthy. In modern draw texturising, the drawing and texturising operations are combined in a continuous or simultaneous manner, thus giving a uniform treatment of drawing and texturising to the resultant texturised yarns. Thus, instead of three processes, namely, extrusions or spinning, draw twist or draw winding and texturising, two processes are adapted, namely, spinning and draw texturising of UDY.

There are some technical difficulties in using UDYs, such as

(a) The UDY has a poor shelf life.

(b) The UDY packages are unstable, which causes the variation in tension and filament.

(c) If UDY reaches zero tenacity temperature, that is, the temperature at which the filament completely loses its strength, it becomes useless for texturising.

(d) The UDY is very sensitive to moisture and temperature changes.

(e) The UDY poses difficulties on texturising machine.

To overcome the above drawbacks of UDY, a compromise was reached in using a POY. Thus, there are now more alternatives for texturising the yarn besides the conventional three-stage process which are as follows:

(a) Spin draw and texturising, or
(b) Spin part draw and draw texturising of POY or
(c) Spin and draw texturising with UDY.

In the first alternative, spinning and drawing is done in one continuous operation. Filaments are extruded at a normal spinning speed of 1000–1300 m/min and then passed through a zone defined by a pair of godet rollers, the output roller of which runs at a speed of 2500–4000 m/min. The fully drawn yarn is then wound up and the wound packages are then passed on to the conventional texturising process.

In the second alternative, drawing process is broken into two parts. Yarns are partially drawn after spinning and the rest of drawing is done at texturising stage. This is the most common method of draw texturising followed all over the world. The draw texturising can be carried out in two ways. They are sequential draw texturising and simultaneous draw texturising. Among the two, the simultaneous method has attained greater importance. It is more economical and has the additional advantages that the draw heat is fully utilised for heating the yarn. All the friction texturising of polyester is carried out by this method and it is increasingly being used for texturising coarse and medium deniers of polyamide. In case of polyester, POYs with DR of 1.5–1.8 have proved to be specially suited for friction texturising by the simultaneous method. The sequential method is often preferred for hosiery yarn of low spinning speeds or crimped polyamide yarn.

Sequential draw texturising: Sequential texturising means drawing and false-twisting in a two-step sequential process. In the first step, the yarn is drawn to achieve a good orientation, and in the second step it is false-twisted. It has the advantages of lower draw tension of LOY compared with POY resulting in lower process tensions resulting in lower surge.

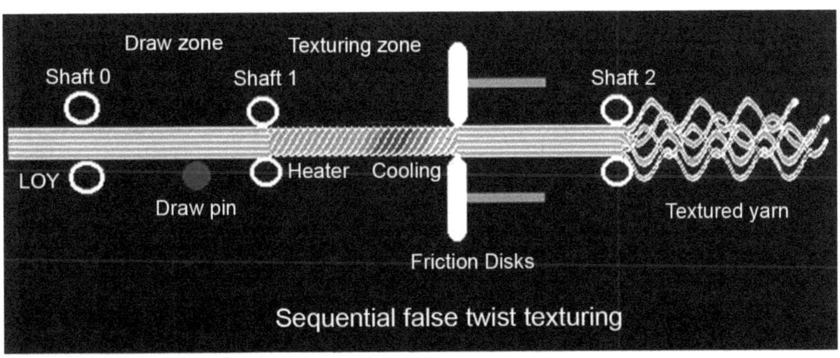

Fig. 2.4: Sequential false-twisting

Shaft 0 is the input feeding device for LOY. The drawing occurs between shaft 0 and shaft 1. In this drawing zone, about 90% of the total amount of drawing will take place. Polyester will need to be drawn over a fixed hot pin of about 65 mm diameter and 360° wrap to achieve good texturising results. Temperatures have to be around 90°C–100°C. Polyamide can be drawn over a fixed cold ceramic pin, about 5 mm diameter with 360° yarn wrap or no pin at all. The pin helps to localise the draw point. However, this process is generally used for polyamide POY as the conventional spinning is obsolete.

The remaining amount of draw is applied in the texturising zone, giving the yarn the necessary tension for a stable process. The twisting is done with a friction device, such as a set of rotating friction discs. After shaft 1 there is a heater, which heats the yarn to a temperature where it can be thermo-set. Adjoining the heater is a cooling plate, which must cool the yarn to a substantially lower temperature to permanently thermo-set the twist. As the yarn is released from shaft 2, each filament tries to assume three-dimensional helix formations it was set in. The result is a voluminous bulk yarn.

Sequential texturising can also be done on double-heater set yarn machines.

It has been known throughout the recent expansion of the practice of false-twist texturising that the conventional false-twist texturising process could be combined with the conventional drawing process performed by the producers. The known procedures for combining drawing and false-twist texturising include both sequential and concurrent performances of the two processes. Sequential performance involves the utilisation of conventional drawing equipment in series upstream for the false-twist texturising equipment so that the undrawn feed yarn is first drawn and then false-twist texturised both at compatible speeds. The sequential performance of drawing and false-twist texturising, effectively in two zones on the same machine, produces yarn which is comparable to conventionally pre-drawn texturised yarn. From an overall process standpoint, the sequentially combined procedures decrease the yarn handling required in comparison with the conventional procedures. On the other hand, the efficiency of the conventional drawing procedure is reduced when sequentially combined with false-twist texturising due to the lower speeds at which the drawing must be performed to make it compatible with the false-twist texturising.

The input material is undrawn or partially drawn supply package, so drawing is essential. It is done by using draw rolls and hot pin in draw heater. The yarn then passes on to the input feed roll; there, the rest of the process is similar to false-twist texturising method.

While using LOY, it is essential to verify the shelf-life. Older the yarn, it is difficult to texture.

Simultaneous draw texturising: Simultaneous texturising means drawing and false-twisting the yarn in a single-step process. In case of simultaneous draw texturising, instead of undrawn or drawn supply package, POY supply packages are used. The passage of yarn is much similar to that of falsc-twist tcxturising method. LOY cannot be used on simultaneous texturising machine.

Concurrently combined drawing and false-twist texturising involves drawing the yarn essentially simultaneously with false-twisting. The equipment utilised thus provides but one treatment zone as distinguished from the two treatment zones of the sequential procedure, in which single zone the undrawn/partially drawn feed yarn is both drawn and false-twist texturised. The improved dyeing uniformity of the resultant product, compared with similar yarn produced from previously fully drawn yarn, is attributed to the lower rate at which drawing takes place and to the concomitant increased time for heating the yarn. From an overall process standpoint, the same advantage of a decrease in yarn handling is obtained compared with the conventional separate drawing and texturising. However, a severe operating difficulty is introduced which is not present in either conventional false-twist texturising or two-zone draw texturising. This difficulty is that of undrawn feed yarn rupturing in start-up. A system to solve this problem was patented recently under US 3816994.

Turbo-Duo-Twist

This method of texturising is gaining more importance because of the noiseless mechanism and very high production rate. This method does not require the use of complicated and costly false-twist mechanisms. It is based on the principle of twisting two or several yarns together with simultaneous heat-setting and consecutive separation of these yarns and winding them on to take up ~~the~~ packages. The yarn coming from the supply packages passes through the hook of the yarn guide and goes along the channel of additional bobbin to the point of assembly and twisted with a second yarn supply coming from the package. The preliminary twisting of the yarns is affected during threading when a certain length of the yarn is wound on bobbin. The number of yarn coils on the bobbin must correspond to the number of twist inserted into the twisted yarn; as the yarn is unwound from the bobbin in the axial direction, it will obtain the number of twists equal to the number of turns it makes around the bobbin. The yarn end of the bobbin is knotted with the end of the yarn coming from the package. Upon yarn unwinding from complementary bobbin, the process is carried on continuously. The yarns which have been twisted together pass the guide, enter heater box and are brought by the guide rollers to yarn separating device. The separated yarns are wound on to take up ~~the~~ packages.

The yarn can be given either S- or Z-twist depending on the direction of twist inserted into the yarn at its winding on to the complementary bobbin. Yarn separating mechanism is the most important unit of the machine affecting texturising by separating two yarns twisted together.

This method enables texturising of finer denier (15–70 denier) at a very high rate of production of 350 m/min in a practically noiseless machine as they have no false-twisting mechanism and driving belts. Moreover, this method of texturising allows considerable reduction in electrical energy.

The twist texturised yarns find use in tricot fabrics, hosiery and all types of knitted outerwear.

2.1.2 Non-Torque Crimping

2.1.2.1 Stuffer Box Method

The stuffer box method depends on the ability of the filament to acquire a permanent set by heat. The process of texturising by the stuffer box method is based on the principle of heat-setting filaments held in a confined place in compressed state and then withdrawing them in their crimped (curled) form. The chamber in which the filaments are stuffed is known as 'stuffer box'.

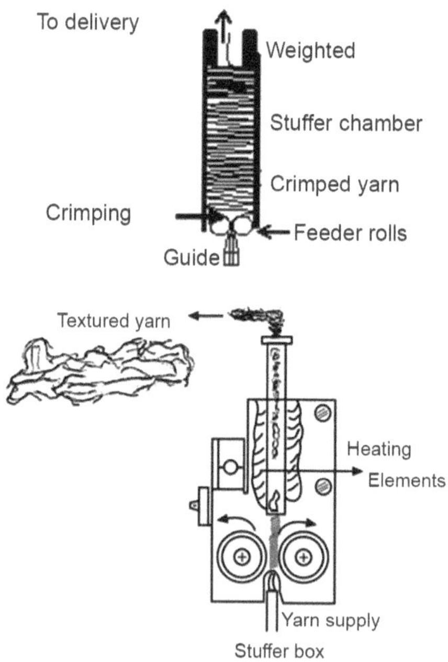

Fig. 2.5: Stuffer Box Texturising

A pair of feed rollers supply the filament yarn with producer's twist amounting to 0.5 turns/inch into an electrically heated tube. The yarn is packed within the stuffer box. The individual filaments are sharply bent by the compression created by yarn packing. This produces crimp in the yarn. The crimp is simultaneously heat-set by the electric heating element while the yarn is in the compressed condition. The electric heating element surrounds the tube as shown in Figure 2.5. The crimping unit consists of two feed rolls and a brass tube stuffer box. By compressing the yarn into the heated stuffer box, the individual filaments are caused to fold or bend at a sharp angle, while being simultaneously set by a heating device. The thermoplastic feed yarn is positively fed by two feed rollers into the heated tube stuffed box. On the output side, the yarn is passed through a weighed hollow tube or slug that impedes the progress of the crimped yarn travelling up the tube, thereby causing the yarn to back up inside the stuffer box tube. The feed rollers at the same time keep delivering fresh yarn against the backed-up aggregate in the tube. This operation causes the filaments of the yarn to stack up in bent or crimped form. The crimp resembles the saw tooth configuration. While the feed yarn is being delivered to the stuffer box, the crimped yarn is removed at a predetermined rate and wound on to the take-up package, and the aggregate in the tube move up and is heat-set at a required temperature in the stuffer box, which is jacketed by the heaters. The yarn is oiled as it emerges from the weighted tube, and it is then coned.

The ratio of feed to take-up and the weight of the tube determine the crimp amplitude and frequency. The temperature of setting and the feed yarn characteristics are the additional factors that determine the crimp quality of the bulked yarn. The tension applied during the heating and cooling cycle affects the degree of stretch and recovery characteristics that resultant yarn possesses.

The stuffer box produces suitable yarn for carpet industry as the yarns made by this process have a great degree of bulk and some stretch.

2.1.2.2 Edge Crimping Method

In this method of texturising, the thermoplastic yarns in a heated and stretched condition are drawn over a crimping edge and cooled. The crimp development in a filament is caused by the bilateral compressional and extensional strain deformations induced in the heated filament as it is being pulled over a blunt knife edge. The molecules of the filaments near the scraping edge reorient parallel to the edge. The molecules of the filaments away from the edge become parallel to the axis. This change in orientation of the molecules will shorten the filaments on the scraped side and therefore induce a crimp.

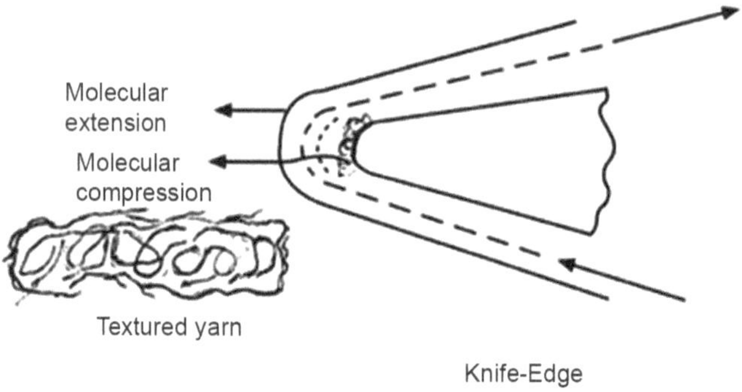

Fig. 2.6: Edge crimp texturising

Bending of the filaments over the edge produces compression on the lower side and extension on the upper side. As the stresses decay, the filaments will be set in a crimped state. The above two effects will cause the mono- or multifilament yarn to coil up on release of tension and each filament of the edge-crimped yarn will take the form of a torque-free reversing helix.

The thermoplastic filament yarn is unwound from the supply package and is passed through tension guiding rod and drum to hot knife edge, which is electrically heated. After running two or three times around the knife edge and the drum, the yarn is guided by the guide rod and wound on to a take up ~~the~~ package by means of the winding drum. The angle formed by the ascending and descending branches of the yarn from the knife edge may be changed by displacing a moveable rod below the knife edge. The angle should be above 20°. The stretch yarns made by the above method have a high degree of elasticity and are often used to make hosiery yarns. *Agilon* is the trade name of yarns produced by this method.

2.1.2.3 Gear Crimping Method

In this texturising method, the thermoplastic yarn is fed through the closely meshing teeth of two gears. The gear head is heated so that the yarn takes on the shape of the gear teeth and the crinkle produced is permanent. The yarn from the supply package passes through tension arrangement, guide hook, feed roller, round the hot braking stud and is subjected to texturising between the cold rotating gears. Furthermore, the twisted yarn passes through the lappet thread guide and is wound on to take up ~~the~~ package.

The frequency and the amplitude of the crinkle or crimp can be varied over a wide range, depending on the types of gears used. Mirlon gear compressing unit utilises saw toothed gears instead of plain gears for texturising. The gear crimped yarns are used in a variety of end-uses' applications, such as ladies and children knitted outerwear, sweaters and ladies blouses. The texturising method of Mirlon allows the production of yarns with a linear density of 2–100 Tex at a linear speed of up to 200 m/min. The number of curls/crimps can be changed from 7 to 20/cm depending on the size of the gear tooth. This method is widely used in staple fibre draw line to introduce crimp before the tow is cut into staple fibres.

2.1.2.4 Knit-de-Knit Method

In this method of texturising, the flat yarn is first knit into a 2-inch diameter hose-leg, heat-set in an autoclave and then unraveled and wound onto a final package. This texturising method produces a crinkle yarn. The crimp frequency and shape can be varied by varying the needle gauge, knitting structure and so on. The fabrics produced by this method have pronounced sparkling boucle texture, excellent stretch and recovery from stretch and full hand.

The stretch yarn coming from the supply package runs around the guiding rods and enters the head of the knitting machine and the obtained fabric is subjected to stabilisation in the box, after which it is wound on the balls or packed in the boxes. The stabilised fabric is unraveled on a high-speed machine and the yarn is wound on the take-up package. Usually, this machine is provided with an auxiliary device to make the yarn more bulky and fluffy.

The production of the crimped yarn by this method has proved highly efficient and high-speed circular knitters have been developed for this purpose. The production capacity of the unraveling machine is also high as they work at speeds up to 700 m/min.

One of the advantages of the knit-de-knit method of texturising is the possibility of varying the yarn extensibility, crimpiness and so on, by using various knitting heads, different depths of sinking, needles of different shapes and various initial yarns.

The texturised yarns produced by the knit-de-knit method are widely used for hosiery goods, knitted outwear, upholstery and so on.

2.1.2.5 Bicomponent Fibre-Texturised Yarn

The bicomponent fibres can be defined as 'extruding two polymers from the same spinneret with both polymers contained within the same filament'. In the bicomponent fibre-texturised yarns, two different polymers with different shrinking properties are used. The side-by-side fibres are generally used as self-crimping fibres. There are several systems used to obtain a self-crimping fibre. One of them is based on different shrinkage characteristics of each component. All commercially available fibres are of this type. There have been attempts to produce self-crimping fibres based on different electrometric properties of the components; however, this type of self-crimping fibre is not commercially used. Some types of side-by-side fibres crimp spontaneously as the drawing tension is removed and others have 'latent' crimp, appearing when certain ambient conditions are obtained.

The nano fibres with side-by-side morphologies are electro spun from polyacrylonitrile and polyurethane using a microfluidic device. Laminar flow of the two polymer solutions through the device results in nanometer-diameter curly nano fibres with bicomponent cross sections. The polyurethane half of the nano fibres can be dissolved in tetrahydrofuran, leaving a 'U'-shaped cross section as seen in Figure 2.7.

Fig. 2.7: Bicomponent fibres

Several factors are crucial to the fibre curvature development: the difference in the shrinkage between the components, the difference between modulus of the components, the overall cross-sectional fibre shape and individual cross-sectional shapes of each component and the thickness of the fibre.

2.1.3 Texturising Techniques Not based on the Thermoplastic Nature of the Filament

2.1.3.1 Bulk Yarns

These yarns have almost no stretch but are bulky in appearance. These yarns are more predominantly used with advantage in carpet and upholstery trade, where their high bulk and rough texture are exploited. Acrylic and nylons are extensively used in this trade. Bulked yarns can be obtained by two methods.

(a) Bulk developed in
   i. Profile
   ii. Hollow
   iii. Combination of hollow and profile
(b) Bicomponent
   i. Bulk developed in yarn
   ii. Air bulking (air texturising)
   iii. IIi-bulk yarn
   iv Composite yarn

Mainly, gear crimping, stuffer box and air jet methods are used to produce these yarns of increased bulk but minimum stretch. These processes can increase bulk from 100% to 300%. They are used as carpet yarns and in sweater fabrics. The resultant fabrics are usually soft with some degree of bulk and warmth but are light in weight.

Air Jet Method

Air jet texturising works by mechanical interlocking, not by heat-setting. It can, therefore, be applied to any continuous filament yarn, including rayon, glass and the new high-performance fibres, as well as nylon, polyester and polypropylene. The method was invented by DuPont in the 1950s. The basis of the method is that the yarn is overfed into a compressed air jet stream, so that the loops are forced out of the yarn. Because the loops need to be locked into the yarn, this can be achieved by twisting the yarn at the take-up. The alternative, which is the current practice, is to design the jets and the yarn path so that there is sufficient entanglement in the core of the yarn to stabilise the loops. Air jets are also added to false-twist texturising machines which enable yarns of a different character known as intermingled to be used directly in weaving.

In air texturising machine, continuous fibres (POY/FDY) such as nylon and polyester are fed into an air texturising nozzle with a feed roller set at arbitrary speed. It interlaces that yarn with air or blends them using turbulence generated by compressed air inside the nozzle, and then produces yarn known as air-texturised yarns (ATYs).

In this method of texturising, the yarn is led through the turbulent region of an air jet at a rate faster than it is drawn off on the far side of the jet. In the jet, the yarn structure is opened, loops are formed and the structure is closed again. Some loops are locked inside and others are locked on the surface of the yarn. The basic principle of this method is that the air stream creates a turbulence that causes the formation of random loops in overfed individual filaments. In this process, the yarn contracts in length, and as it emerges, the loops are locked in place to impart bulk to the yarn. The yarn thus produced has an appearance like a staple yarn, but possesses high bulk, greater covering power, reduced opacity and a warmer hand compared with flat continuous filament yarn. The texturised yarn produced by this method is designated as Taslan.

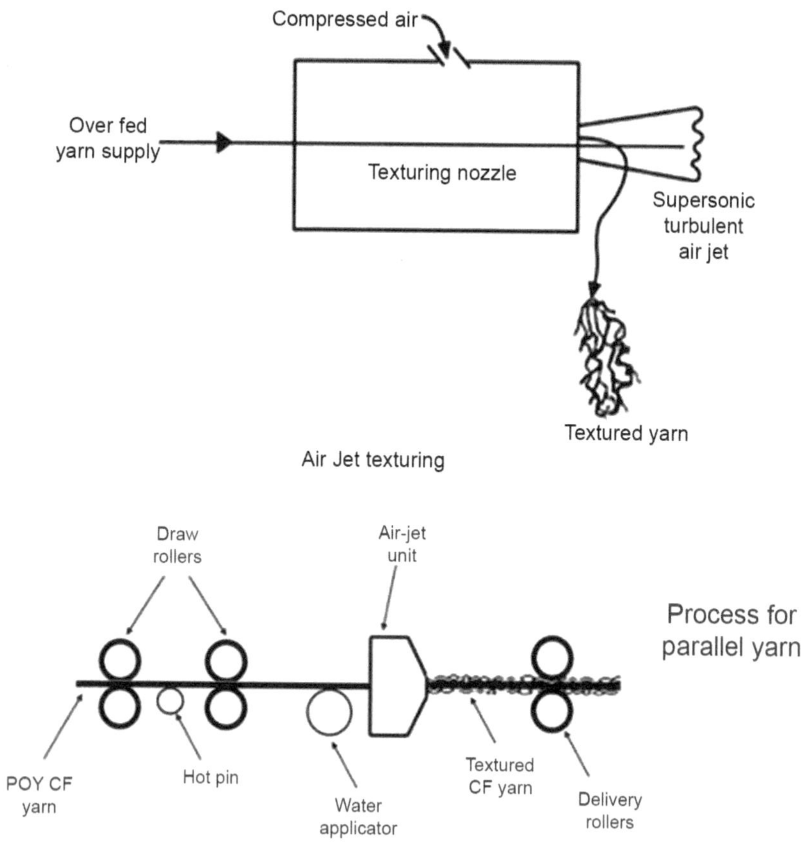

Fig. 2.8: Air jet texturising

The yarns produced by the air jet texturising are totally different structures in that they much more closely simulate spun yarn structures. Whereas the bulkiness of the stretch yarns decreases with the degree of the tension imposed on them, the form of the air-texturised yarns can be made to remain virtually unchanged at loads corresponding to those normally imposed in fabric production and during wear. This is due to the locked-in entangled loop structure attributed to the air-jet-texturised yarns.

ATYs can offer entirely different look and hand to any fabric. It can be used in a very broad range of fabrics from simple lightweight scrim to very heavy-duty soft luggage fabrics from lightweight swim suit to heavyweight parkas. Several filaments can be combined to suit the end-use requirements. Low denier ATYs are costly whereas high denier yarns can compete comfortably with spun yarns.

When compared with DTYs, ATYs are more bulky with zillions of small loops. All loops in ATYs contribute to its bulkiness and loft feel of the fabric.

ATY is very bulky with permanent crimps and loops. Interlacing of the filaments in the jet can cause the loops to be locked into the yarn, so that twist is unnecessary. Loop frequency, loop dimensions, loop stability and physical bulk are the important characteristics of the air-jet-texturised yarns.

Nylon ATY is often used in sportswear, umbrellas and bags. Spun-like nylon yarn is mass-produced in Taiwan and South Korea. Fabric woven with this yarn has been utilised in famous sports brands and is well-known all over the world. In the polyester field, ATY is being used in the interior of the vehicles such as car sheeting and as materials for interior decorating by European, Japanese and other global automobile and interior material manufacturers in recent years. As described, ATY is extremely versatile due to its variety of forms and characteristics. Demands for ATY are expanding due to the development of products that differ from traditional DTY (DTY).

Air-entangling and co-mingling should not be confused with air jet texturising. Air-entangling of the filaments to keep them together is more economical than inserting twist. The air jets are used to entangle yarn filaments in programmed and repeating areas called nodes or knot (tack) points. This makes it easier to process the yarn in preparation and fabric forming steps. It also aids in co-mingling or tying down plied yarns. The frequency of tack points, that is, interlacing/intermingling points, is described as points per metre of yarn. The key to successfully weaving the texturised yarn is co-mingling unless the yarn is twisted. Intermingled points should remain in the yarn through weaving but should come out in fabric preparation processes prior to dyeing and finishing of the fabric. If they are not removed, then pin holes will be noticed in the fabric which is unacceptable.

Principle of producing parallel ATYs: One or more ends of filament yarns are overfed at a constant rate to a special air jet which blows, depending on the overfeed between the inlet and outlet feeds, the yarns into a continuous string of smaller and larger loops. The filaments in the centre get braided by tangling action. The individual filaments are compacted by air stream which stabilises the loops. Water is added in front of the jet to the yarn, so that there is less yarn to jet friction inside the jet and the formation of the loops becomes more efficient. The water is blown off the yarn at the jet exit and the yarn is basically dry when it is wound on to a package.

Figure 2.9 represents the simplest way to produce ATYs. As POYs became widely available, a hot pin and a drawing zone were added to draw the POY as shown in Figure 2.10.

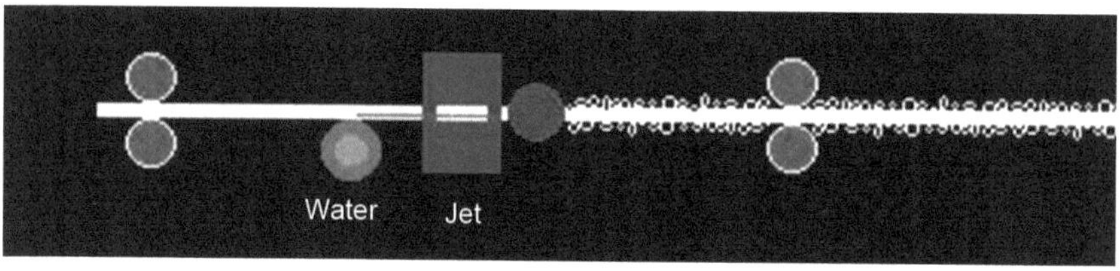

Fig. 2.9: Simplest way to produce air-texturised yarn

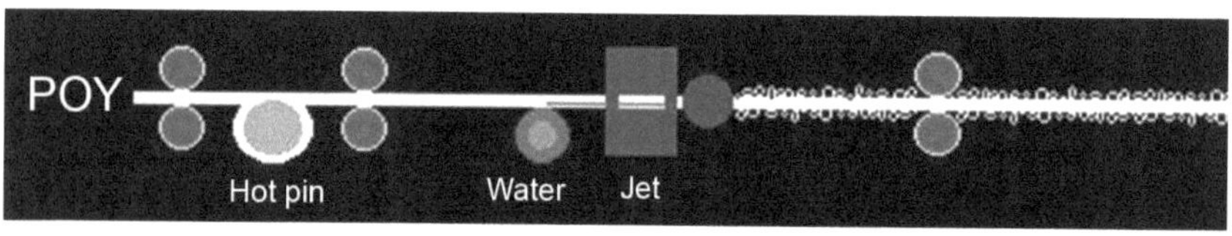

Fig. 2.10: Hot pin and drawing zone added to air texturising

Principle of producing core effect in ATYs: The core yarn is overfed to the jet with less overfeed than the effect yarn. Jet manufacturers are providing different jets for high and low overfeeds. The final denier also influences the jet design.

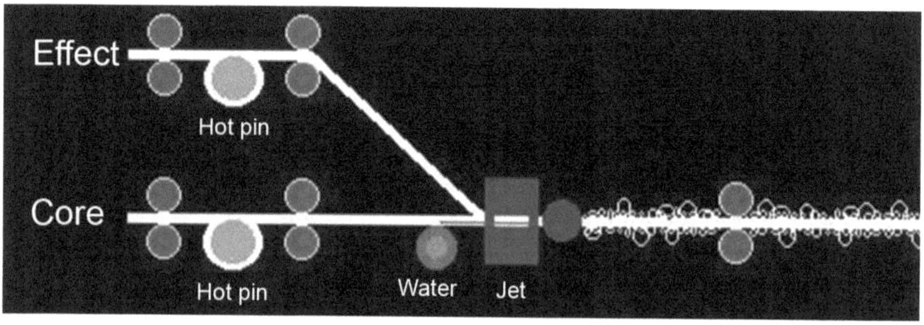

Fig. 2.11: Core effect in air texturising

Loop size and loop stability matter greatly, especially for fine denier yarns. Smaller loops are more stable and provide better volume than larger loops. To increase the stability and to shrink the loops, ATYs can be heat-treated in a setting heater as shown in Figure 2.12.

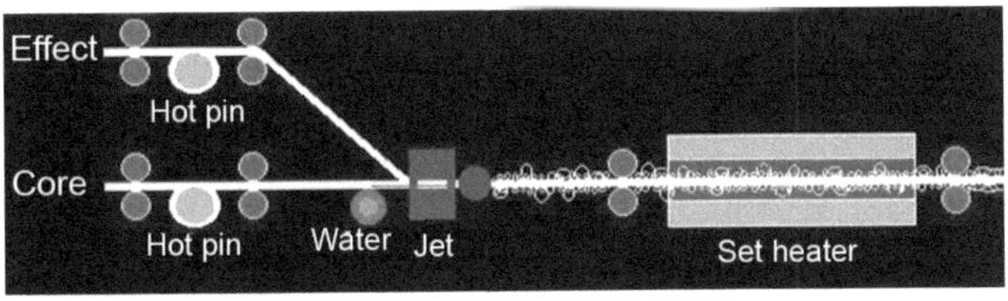

Fig. 2.12: Heat-setting added to air texturising

Specially stabilised ATYs can be produced with an additional drawing zone between the jet and the heater. Here long loops are stretched to a point where they are not a loose loop anymore and the rest of the loops are drawn to a smaller size. This increases yarn stability and process ability.

Figures 2.13 and 2.14 shows the process patented by Barmag where the drawing of polyester can also be done by using hot draw godets. Hot pins are stationary and the yarn is drawn by wrapping the pin up to 360° angle. Hot godets are wrapped several times to allow the yarn to absorb the heat of the godet. The first godet is normally heated above the glass transition point of polyester to allow optimum drawing conditions. The second godet, if heated, is used to anneal the fibre.

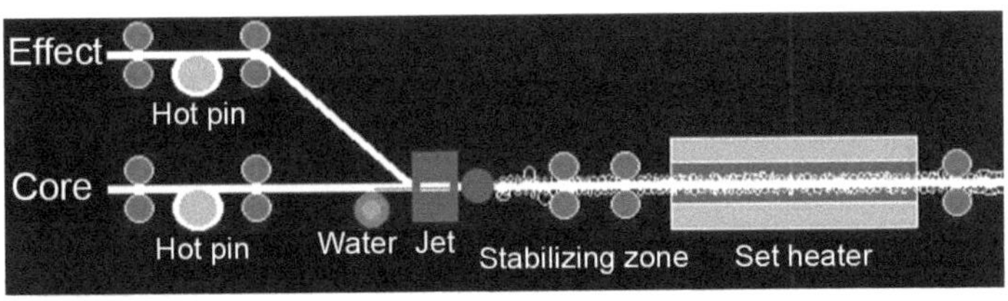

Fig. 2.13: Stabilising zone added to air texturising

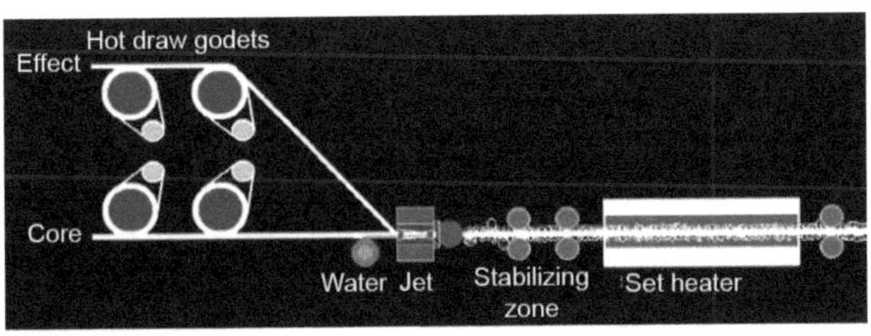

Fig. 2.14: Hot draw godets added to air texturising

2.1.3.1 Compound or Composite Yarns

Compound or composite yarns are yarn structures consisting of at least two strands, one forming the centre axis or core of the yarn and the other forming the covering or wrap of the yarn. One of the strands can be filament yarn and the other a spun yarn or both can be filament yarns in order for the classification to be a compound or composite filament yarn. These yarns are generally available in the same size ranges as filament and spun yarns. In general, there are two types of compound yarns, covered and core-spun. In the covered filament yarns, the wrap is the yarn which can be either a filament yarn or a spun yarn. The core might be elastomeric, such as rubber or spandex, or hard, such as polyester or nylon. The covered yarns may have either a single covering or a double covering. The second covering is usually twisted in the opposite direction from the first covering. Core-spun yarn has a hard or elastic yarn core along with a sheath of staples on the yarn surface. A very common core-spun yarn has spandex in the core and cotton staple twisted around the surface of the yarn. Core-spun yarns can be of three types – filament core/staple wrap, staple core/filament wrap and staple core/staple wrap.

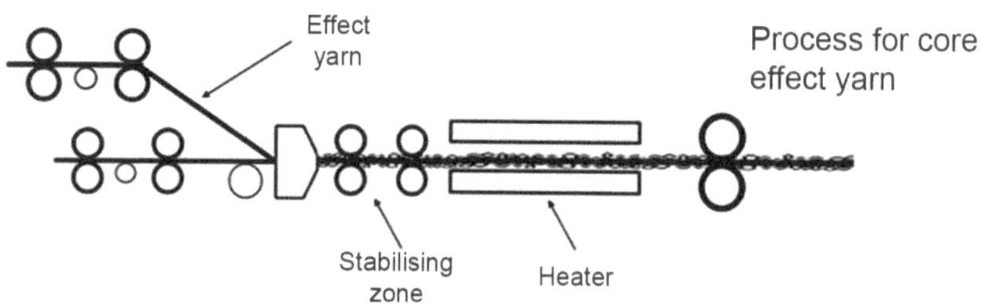

Fig. 2.15: Process for core effect yarn

The core could also be a hard filament yarn, texturised or flat, or a hard spun yarn. The supply package of spandex is surface driven at a controlled rate. The yarn passes from the package drive-roll to a driven feed roll called a star wheel. The yarn then passes through two hollow spindles on which spools of the covering yarn are mounted, and to the driven take-up wheels or to nip rolls. For double-covered yarns, spools revolve at high speed in the opposite directions, wrapping the hard-yarn around the stretched core of spandex with sufficient wraps per unit length to permit the yarn to have the desired stretch. The bottom cover yarn normally controls the stretch, while the top cover serves to balance and give a smooth appearance to the surface of the covered yarn. With a bottom and top wrapped in the opposite directions, 'S' and 'Z', what is produced is a torque-free double-covered yarn with definite stretch limits. For single covered yarns, a single spool of covering yarn is mounted on the bottom spindle to wrap the stretched core of spandex with sufficient wraps per unit length to give the desired yarn aesthetics and protection to the core. The covered yarn is allowed to relax partially above the take-up star wheel and is then wound onto the take-up package. The surface speed of the take-up package drive-roll, relative to that of the take-up star wheels, governs the relaxation of the covered yarn. The total draft is chosen to give the best balance between the weight and cost of the spandex, machine productivity and the appearance of the covered yarn for the power and stretch required. In practice, lower drafts are generally used to give the best balance of the yarn properties. Excessive drafting can cause broken filaments in the spandex core. A covered yarn with such a core will contain slubs and show variations in power. The covered yarns of low denier per filament (dpf) will produce smoother covered yarns and better coverage of the core. With double-covered yarns, the stretch is controlled by the number of wraps of the bottom cover per unit length. The balance of the yarn is controlled by the number of wraps of the top cover.

During the covering process, the yarn coils produced by the covering yarns are spaced apart. However, upon relaxing the yarn, the coils become closer and more compact. As previously mentioned, with double-covered yarns the stretch is controlled by the number of wraps of the bottom cover per unit length. The balance of the yarn is controlled by the number of wraps of the top cover.

Filament core/staple wrap core-spun yarns have filaments in the centre of the yarn which are completely surrounded by a wrap of spun yarn. The wrap is sometimes referred to as a sheath because it is composed of untwisted staple fibres. This type of yarn has applications in apparel, household and technical areas. When plied, polyester core/cotton wrap yarns make sewing threads marketed as 'cotton covered polyester'. Some protective clothing and flame-resistant curtains are composed of glass-filament core, surrounded with aramid, aramid blends or PBI. Other yarns may have carbon in the core for these same applications. Some protective gloves may have steel wire surrounded by aramid blends. When cotton is the sheath, it allows for good dyeing, good hand and protection for the elastomeric. Yarns of this type provide stretch in denim, knit and hosiery products. Staple core/filament wrap core-spun yarns have a core of untwisted, parallel staple strand in the core of the yarn and a filament wrapped around the core. This type of yarn can be a good source for using wastes, which can be placed in the yarn core. Because twist is not inserted, the most expensive aspect of producing a spun yarn is eliminated. The majority of the development and production of staple core/staple wrap yarns has focused on the use of polyester staple for the core and cotton for the wrap.

2.1.3.2 Stretch Yarns

Stretch yarns are characterised by high extensibility (say about 300%–500%), rapid recovery and moderate bulk per unit mass. These yarns are suitable for socks, hosiery, slacks and other garments where snug fitting and more stretch is required. Polyamides dominate in this region, and the weft knitted garment manufactured out of stretch nylon dominates ladies foundation garment trade.

False-twist, knit-de-knit method and edge crimping method are used to produce this type of yarns having 300%–500% elongation. In the first method, the yarn is twisted, heat-set and untwisted to get a coiled yarn. In the second method, the filaments are knit into a hose, heat-set and then deknitted after cooling, whereas in the third method the hot filaments are drawn over a knife like edge, which flattens one side and causes the yarn to curl. They are used in swimsuits, lingerie, stockings and one-sized garments where a form-fitting resilience without pressure is required.

2.1.3.3 Modified Stretch Yarns

They are between the above two in stretch properties (about 10%–15%). They are made in basically the same manner as stretch yarns except for a final step of heat-setting the yarn after the untwisting step. The heat-setting is given under certain conditions which modify the characteristics of the yarn. The primary intention of this modification is to make them suitable for some end uses like stretch-to-fit applications and hence to reduce the stretch ability to some degree and thereby increase the bulk and giving an excellent appearance and hand.

Stretch-yarns are generally converted into modified stretch yarns by any of the following methods:

(a) Overfeeding them up to about 30% or more through a second heating zone.

(b) Heat-setting or stabilising soft wound package of stretch yarns under suitable conditions either in autoclave, that is, by steam stabilisation or in a yarn dyeing operation.

The yarns produced by the first method are generally called as 'modified stretch', 'stabilised' or 'post treated' yarns and those produced by the second method are called 'the set texturised yarns'.

The set texturised yarns are being used extensively, particularly for knitted outwear applications. Polyesters dominate in the set yarn consumption over nylon as they are used for warp knit fabrics for tailored men's wear items as well as for circular double knits for women's dress and suit. The advantages of set polyester yarns are as follows:

The bulk in set yarns being stable, fabrics can be knitted from these yarns much closer to the finished size than those knitted from stretch yarns, and Stabilising treatments during finishing are not generally required when the set yarns are used.

2.1.3.4 Tangled/Interlaced Yarns

Interlacing is done to provide cohesion to synthetic filaments, which was earlier done only by twisting. A continuous yarn running under a defined tension through an air jet can be interlaced if a perpendicular or nearly perpendicular high-pressure air stream is applied to the yarn. The air stream creates turbulence splitting the yarn bundle and then forcing individual filaments together, which creates a kind of braiding effect on the yarn. This increases the cohesion between the filaments. Such DTYs can now be used in weaving directly without a need of further twisting.

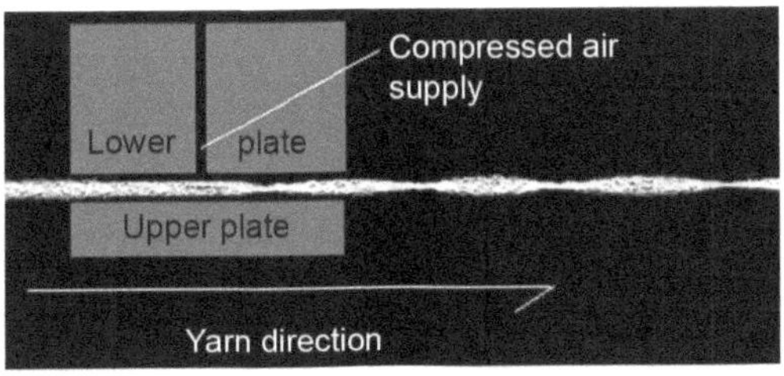

Fig. 2.16: Tangled/interlaced yarn production

A multifilament yarn is fed through the tangling jet. The perpendicular compressed air stream will split the filament bundle. Due to a very high dynamic force, the filaments collapse into the filament bundle where they now entangle. The entangled yarn is characterised by having tangle knots at very regular intervals. This evenness for subsequent process is very important. The density of the tangle knot is controlled by air pressure and yarn tension. The evenness of the knots is a result of the evenness of the yarn, the yarn tension and the compressed air pressure. The design of the air jet and the angle of the yarn path in and out of the jet are mainly responsible for the frequency of the knots and the actual air consumption. The shape of the yarn channel and the size of the orifices of the compressed air channel vary between the manufacturers of the interlacing air jets.

Figure 2.17 shows different positions where a tangle jet can be located in a machine.

- The first row shows the ideal situation, where the jet is located between a set of feed shafts, allowing controlling the tangle tension of the process.

- The middle row demonstrates where a jet can be located on older equipment. The advantage of this location is easy access and supervision of the process, but a disadvantage is the fact that the tangle tension is governing the setting tension.
- The last row shows the location of the jet after the second heater and before the last feed shaft. The disadvantage is inaccessibility and less control. This location was favoured for long as it would allow producing a uniform set of yarn before that yarn was tangled.

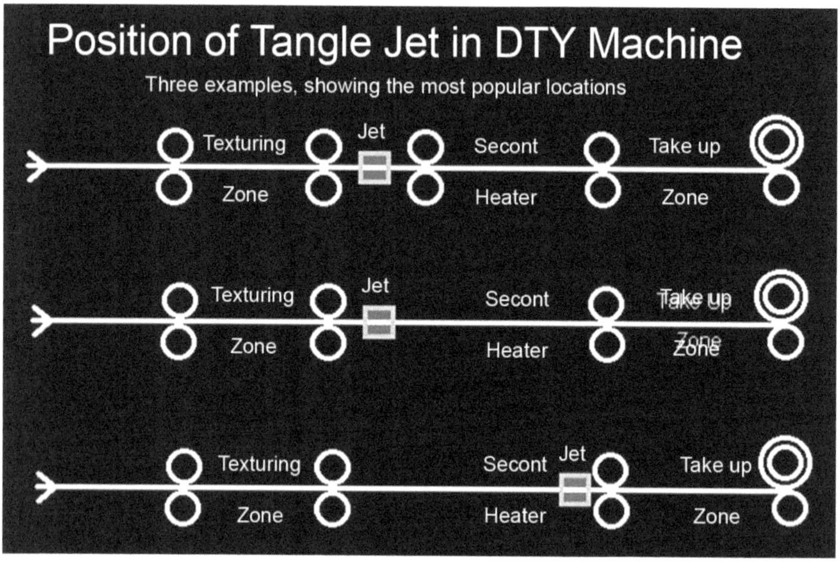

Fig. 2.17: Position of tangle jet in DTY machine

Synthetic filament yarns, regardless of POY, FOY or DTY, can be interlaced. The lower the denier ~~from~~ of the filament and more the filaments, it is easy to achieve a cost-effective tangle result. POY is interlaced for better package formation and better texturising results, for example, less filament breaks. FDY is interlaced for better package formation and for better warping and weaving results.

2.1.4 Texturising Techniques – Natural and/or Man-made Fibres

2.1.4.1 Slack Mercerisation of Yarns

In normal mercerisation of cotton yarn, the mercerisation is done by keeping the skein in a stretched condition to get proper orientation leading to lustre and higher strength. If the mercerisation is done in a slack form, the yarns shrink heavily producing a crimped-type effect. The shrinkage shall be asymmetrical as there is no control. The crimps shall be high at some places and low at some other places, but there shall be crimp at all places.

2.1.4.1 *Crimping and Crosslinking*

Texturising of the blended yarn faces new challenges due to the complex behaviour of the bicomponent yarn. The setting of deformation in the polyester component requires heat-setting around 200°C, while the setting of the cellulosic component is accomplished through crosslinking, by conventional pad-dry-cure process, at 140°C–160°C. Accordingly, the texturising of polyester/celluloses becomes a two-step batch process. The texturising of the blended yarns could be of commercial interest, if both the components are set simultaneously. Texturising of the blended polyester/viscose yarns has, therefore, been accomplished via the 'twist-set-detwist' (batch), or the 'false-twist' (continuous) process. The setting of the two components can be done through the following two routes:

(a) Simultaneous setting of both the polyester and the viscose components by rapid curing at 210°C, where polyester is set by thermally induced crystallisation and viscose is set by collapsed crosslinking. Various resins are used such as Finish KVS Liquid New (Sandoz), Textile Resin U-4750 (BA.SF) and DMDHEU.

(b) Simultaneous setting of the components by immersing the yarn in a mixture of crosslinking agents (e.g. epichlorohydrin and hexamethylene diisocyanate) and highly plasticising solvents (e.g. tetrachloroethane and dimethylformamide). In this process, polyester is set by solvent-induced crystallisation and viscose by wet crosslinking.

In the first process, it is needed to bring the crosslinking time and temperature as close to the requirements of polyester heat-setting as possible. This envisages the selection of suitable catalysts with considerably enhanced efficiency. The efficiency of various crosslinking catalysts, such as phase-separation catalysts, self-limiting catalysts, basic aluminum chloride, mixtures of glyeol1ic acid and magnesium chloride, was studied in conjunction with DMDHEU to produce texturised viscose yarns by Kushal Sen, where it was observed that magnesium dihydrogen phosphate and basic aluminum chloride are the most efficient catalysts for producing texturised viscose yarns of high crimp rigidity and good strength. These catalysts were then employed to texturise polyester/viscose blended yarns using DMDHEU. Finish KVS Liquid New (Sandoz) and Textile Resin U-4750 (BASF) have also been used for the texturising of the blended yarns.

Magnesium dihydrogen phosphate is used for continuous texturising of the polyester/viscose yarn, using false-twist technique. It is observed by Pushpa Bajaj et al that magnesium dihydrogen phosphate and basic aluminum chloride are very efficient catalysts in view of the crimp rigidity and strength retained after texturising. It was also observed that hydroxymethane sulphonic acid is the only catalyst whose use enhances the strength after texturising, but its applicability is limited, because the crimp rigidity obtained is rather poor.

2.1.4.3 Thermoplasticisation of cellulose

A plasticisation method for cellulose acetates (CAs) based on the selective grafting of ε-caprolactone (CL) and l-lactide (LACD) was developed by Mariko Yoshioka, Nahoko Hagiwara, Nobuo Shiraishi. The selective grafted products were prepared by ring opening polymerisation in the melt state at 140°C using stannous octoate as catalyst, where CAs with the remaining hydroxyl groups worked as the initiator. Plasticisation of CAs by this selective grafting could solve the problem like bleeding of unreacted monomers and homo-oligomers from the inside of the molded articles to their surface. By using adequate reaction conditions, the grafting reaction proceeded rapidly and could be completed within 10–30 min. LACD is grafted more rapidly than CL onto CAs, producing relatively rigid and brittle products in the earlier stages and elastomer-like ones in the later stages. Transparent amorphous molded articles were obtained depending on the reaction conditions. The analysis of the structure of the grafted side chains by means of high-resolution NMR spectroscopy showed that although the grafted side chains are composed of large amounts of ε-oxycaproyl or lactidyl block polymer portions depending on the reaction conditions, a large amount of randomly polymerised parts coexist in the grafted chains, which confer high thermoplasticity, elasticity and amorphous nature to the grafted products obtained.

2.1.4.4 Sheath–Core Technique

Sheath–core bicomponent fibres are those fibres where one of the components (core) is fully surrounded by the second component (sheath). Adhesion is not always essential for fibre integrity. This structure is employed when it is desirable for the surface to have the property of one of the polymers such as lustre, dyeability or stability, while the core may contribute to strength, reduced cost and the like. A highly contoured interface between the sheath and core can lead to mechanical interlocking that may be desirable in the absence of good adhesion.

The most common way of production of sheath–core fibres is a technique where two polymer liquids are separately led to a position very close to the spinneret orifices and then extruded in sheath–core form. In the case of concentric fibres, the orifice supplying the 'core' polymer is in the center of the spinning orifice outlet and the flow conditions of the core polymer fluid are strictly controlled to maintain the concentricity of both components when spinning. Eccentric fibre production is based on several approaches: eccentric positioning of the inner polymer channel and controlling of the supply rates of the two component polymers; introducing a varying element near the supply of the sheath component melt; introducing a stream of single component merging with concentric sheath–core component just before emerging from the orifice and deformation of spun concentric fibre by passing it over a hot edge. Other, rather different techniques to produce sheath–core fibres are coating of spun fibre by passing through another polymer solution and spinning of copolymer into a coagulation bath containing aqueous latex of another polymer. Modifications in spinneret orifices enable one to obtain different shapes of core or/and sheath within a fibre cross section. There is considerable emphasis on surface tensions, viscosities and flow rates of component melts during spinning of these fibres.

A number of new products are being developed using sheath–core techniques, and the majority are still pending for patents. Besides the sheath–core bicomponent fibre used as a crimping fibre, these fibres are widely used as the bonding fibres in non-woven industry. Some examples are discussed here.

Sheath–core techniques using different polymers and cross sections are well known to produce yarn characteristics which may be desirable for specific applications such as tire treads, seat belts and apparel, but which are entirely unsuitable for carpet manufacture and in particular, the manufacture of bulked continuous filament carpet yarn or crimped staple carpet fibres.

Improved filament flexural rigidity could be obtained by using appropriate non-circular shapes in the core and quenching the drawn filaments in 100°C water for between 10 and 15 min. It is possible to extrude two filaments of differing dyeability using a sheath–core system of feeding two molten polymers to a special spinneret. A polyethylene sheath with a core of carbon can be co-extruded with nylon to create a synthetic filament having electrically conductive properties to eliminate static. A bicomponent multifilament that has a pigmented core and an outer sheath of nylon having a high tenacity is suitable for use in seat belts, fishing nets and ropes. The outer nylon sheath protects the manufacturing equipment for seat belts, nets and rope from abrasive additives to the core. By melt spinning a bicomponent sheath–core process using a high IV polyethylene terephthalate core and a nylon sheath composed mainly of polyhexamethylene adipamide (nylon 6/6), it is possible to produce a superior tire cord yarn that has excellent adhesion to rubber

2.1.4.5 Crystallisation and De-crystallisation (Solvent Texturising)

Cotton and viscose rayon stretch yarns have been produced by setting the deformation through covalent crosslinks. The setting of cotton without covalent crosslinks has also been accomplished through a de-crystallisation sequence by A. K. Sengupta. For this purpose, cotton was partly de-crystallised to varying degrees with swelling agents, such as zinc chloride, ethylene diamine and tetra methyl ammonium hydroxide. The de-crystallised cotton yarn was then twisted and re-crystallised by washing out the swelling agent to set the induced deformation.

Simultaneous dyeing and texturising of polyester viscose blended yarn is now possible through union setting of both components at 200°C–220°C. For this purpose, catalysts with considerably enhanced efficiency were selected (e.g. magnesium dihydrogen phosphate and basic aluminum chloride) for setting of viscose component through crosslinking. This is necessary for producing desirable crimp contraction in the blended yarns.

The selection of catalyst suitable for rapid curing was further utilised in developing a unique process of simultaneous dyeing and texturising of polyester and viscose blends. For this purpose, a single bath containing a mixture of disperse and reactive dyes along with dimethylol dihydroxyethylene urea and an acid catalyst was used. Dye fixation and texturising were achieved by baking the yarn on Scragg mini-bulk machine using a heater temperature of 210°C for 2.55 s. However, higher colour yields were obtained by post-alkali treatment for 15 s. Crimp rigidity and the tensile properties of simultaneously dyed and texturised yarns are not affected in any way due to the incorporation of the dyeing step along with the texturising sequence.

2.1.4.6 Interfacial Polymerisation

This is a polymerisation in which two reactive monomers, each dissolved in different solvents that are mutually immiscible, react at the interface between the two solutions. This process can be used for different yarns.

R. E. Whitefield et al. explains the method of stabilising wool using interfacial polymerisation as follows. Poly(hexamethylene sehacamide) formed on the surface of the wool by interfacial poly condensation appears to be grafted to the wool. The ultra-thin film of the polymer is not extracted by good solvents. If the grafting sites on the wool are blocked by acetylation prior to interfacial polymerisation, the surface film is easily extracted with good solvents and the treated wool has no shrink resistance. Application of pre formed poly(hexamethylene sebacamide) from the solution followed by evaporation of the solvent imparts little or no shrink resistance to the wool, and the surface film so formed is easily extracted with good solvents. About 1%–2% polymer can be grafted to the wool surface by interfacial polycondensation; polymer in excess of this amount is only partially grafted. With suitable enzymes to eat away the wool from a treated wool sample, the grafted coatings have been isolated and characterised. The grafting sites in the wool have been shown to be the free amino groups of N-terminal amino acids and amino and hydroxyl groups of side chains in the wool proteins.

2.1.4.6 Differential Shrinkage

The yarns with different shrinkage characters can be combined together while weaving or knitting. After the fabric is formed, we get a puckered effect due to the difference in shrinkage. Similarly, two yarns of extremely different shrinking characters, for example, spandex and cotton, when doubled and twisted together give a crimpy yarn which is highly elastic.

2.1.4.7 Hot Fluid Jets for BCF Yarns

Bulked continuous filament or BCF yarns are another type of the texturised yarn using jets. BCF yarns are used as coarse carpet and upholstery yarns. Their origin lies in DuPont research on jets. The yarn is produced by using fluid texturising with steam to produce texturised nylon and polypropylene carpet yarn. These yarns have a three-dimensional crimp that is preferred for carpet appearance and performance. The steam provides heat to set the bulk in the yarns. In the production of BCF yarn, the yarn is fed through a hot fluid jet. The yarn is then collected in a 'caterpillar' configuration and cooled to stabilise the set, before being taken to the wind-up. The turbulent hot fluid produces an asymmetric shrinkage, which causes filaments to buckle. The buckling effect is set, and a high bulk yarn is produced.

2.2 Advantages of Texturised Yarns

The high stretch and high degree of elasticity in the texturised yarns allow the garment to adjust itself to the shape and dimensions of the body of the wearer, and hence they have replaced the necessity of manufacturing different sizes of garments of knitted goods, for example, socks, and stretch blouses. Garment manufacturers can provide free-size stretch-to-fit items in a small number of sizes for a given wide range of size requirements. This works out economical for the manufacturer and the retailer who needs less space and inventory and also less worry of improper fitting to the wearer.

The fabrics made from the texturised yarn have better drape and are more comfortable as they absorb sweat and moisture readily.

Subdued lustre of the texturised yarn when compared with filament fabric is also an advantage as many consumers prefer subdued lustre than shining lustre. As these yarns are made from filaments of polyamides, polyester or similar thermoplastic synthetics, they retain the properties of parent yarns such as lasting wear and uniform appearance.

A glaring plus point in case of the texturised yarn is that its texture and hand can be varied by engineering the method and condition of texturising.

The fabrics made from the texturised yarns when compared with spun staple yarns give additional plus values such as improved pill resistance, improved crease resistance, greater durability, high abrasion resistance, strength and toughness with bulk. They also give excellent dimensional stability to the final fabric.

They provide excellent non-iron performance achieved by post-cure resin polyester–cotton blends.

It has been found that textiles made from the texturised yarn can be used all the year round under all climatic conditions. They feel warmer than do fabrics constructed from untexturised filaments, because the ordinary untexturised filament dissipates the body heat faster due to its greater area of surface contact with the skin. This warmth without increasing weight of the fabric is the characteristic feature of the texturised yarns.

The texturised yarn fabrics have better shape retention ability than those made from straight filament yarns. This is due to the interfilament frictional properties.

Flexibility is an important property desired in textile yarns and fabrics. A straight filament is less flexible than a crimped, looped or curly filament. The texturised yarns are more flexible compared with the plain filament yarns.

Most of the thermoplastic synthetics have low moisture absorption capacity compared with naturals. However, the air entrapped between the interstices of the texturised yarns imparts good comfort characteristics. As the moisture absorption capacity is low in synthetics, the washing and drying of the fabrics made from the texturised yarns are easy.

2.3 Importance of Different Parts in a Texturising Machine

There are different types of texturising machines, also termed as texturising machines. Some examples are draw texturising machines and air texturising machines. Within draw texturising machines, there are varieties such as polyester polypropylene draw texturising machine, polyamide draw texturising machine, S+Z double-yarn draw texturising machine and air covering draw texturising machine.

In this book, we will deal with the defects observed in false-twist method.

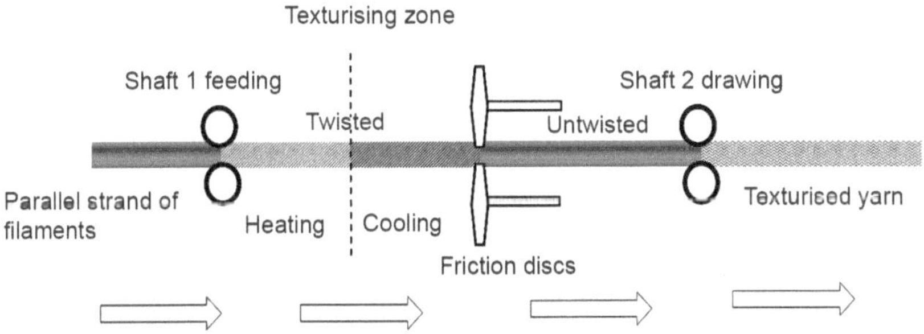

Fig. 2.18: False twisting method of Texturising

Fig. 2.19: V-type draw texturising machine

Let us try to understand the importance of the various parts in a texturising machine.

Friction Discs

Most machines today use friction discs to insert the necessary twist in the yarn for texturising. Many of these machines feed POY yarns which will be drawn at the same time as it is being twisted and texturised. The creel or framework for holding the POY yarn packages is located behind the machine on both the sides. The POY feeder yarns move forward into the first drawing roll, and then flow into the heater zone, followed by the cooling zone. At this point, the yarns go through the friction disc units where twist is inserted into the yarn. The diameter and thickness of the discs, the spacing between the discs, the type of disc material and the speed of rotation all have an effect on the amount of twist that is inserted to the yarn. Below the twisting unit, a second set of rollers deliver the yarn from the discs. A take-up system then winds the texturised yarns onto the tubes to eventually form a large package of the yarn. The speed of the second drawing roller system exceeds the speed of the first drawing roller system. This elongates the yarn and finishes the drawing or aligning of the long-chain molecules within the yarn.

Friction discs insert the necessary false-twist into the running yarn, so that the yarn can get texturised. The discs are commercially available from a large number of vendors such as machinery suppliers and independent spindle manufacturers. The discs will have to conform to a set of design specifications depending on the machine, respectively the friction unit manufacturer. They are not always interchangeable. Specifications that are important to observe are the outside diameter, inner diameter and thickness.

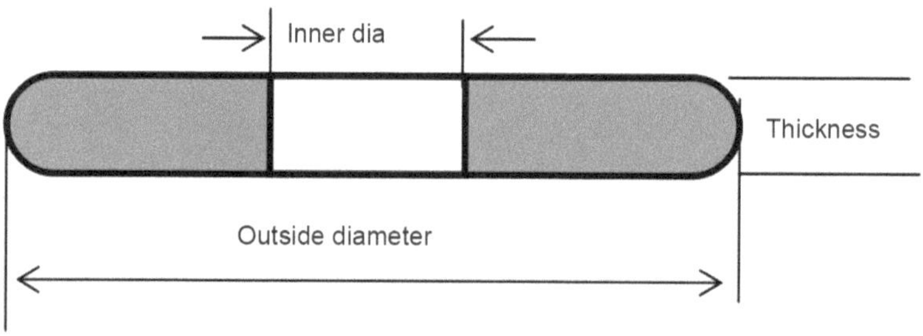

Fig. 2.20: Dimensions of friction disc

In earlier times, when friction texturising machines ran slower, the thickness used to be 4–6 mm. Today, most modern high-speed processes are running on 9-mm-thick discs. Friction discs are expensive, precision made parts. The assembly of discs in a friction unit must be accordingly accurate. The discs are separated by spacers to obtain an exact disc spacing, normally 0.5 mm. Exemptions, such as 0.6–0.75 mm, were made for special needs. Because discs rotate at high speeds, there must be absolutely no interference or touching of the discs, as this will create a high amount of frictional heat, which destroys the expensive discs and can pose, consequently, a real danger to the machine operator.

There are two types of disc materials referred as hard and soft. Hard discs are either metal or ceramic based; soft discs are normally of a plastic/polyurethane construction. Ceramic or polyurethane (PU) materials are the ones which are in touch with the yarn. These discs are known as the friction discs.

Metal Discs: The base material is normally aluminum. The friction surface is coated with a material, which transmits enough friction to the yarn without hopefully abrading the yarn too much. Known coatings are plasma, powder vapor deposition, aluminum oxide nano deposition and diamond nickel coatings.

Ceramic Discs: These discs are normally made of an Al_2O_3 ceramic. They are extremely hard, durable and have a surface, which has proven to change little over time. The earlier discs were quite abrasive, causing a large amount of polymer/spin-finish dust (snow) to accumulate on the machine. This was a continuous housekeeping challenge. The latest generations of ceramic discs have vastly improved surfaces, which still insert enough twist, but reduce the amount of abrasion significantly. In recent tests they have been found to be very close to PU discs.

PU (Polyurethane) Discs: They offer a much higher amount of friction and therefore more twist. As synthetics are normally harder than PU, there is no noticeable abrasion. This results in improvements regarding tenacity and elongation of the texturised yarn. But the softness of the disc also means that its lifetime is limited. This can become a commercial issue. Good discs can last, depending on spin-finish, ambient air temperature, denier and yarn speed, up to 12 months. But even longer lifetimes have been reported. The PU Shore hardness available are 86 and 90 Shore A.

Rubber Discs: Recently, some rubber discs were exhibited. They were then being tested by a texturising machinery producer. It appears now that such discs cannot offer a sustainable advantage over PU discs.

Friction Disc Arrangements

For S- and Z-twists, the discs are stacked differently. The following illustration is an easy reminder of how the discs have to be stacked to obtain S- or Z-twist.

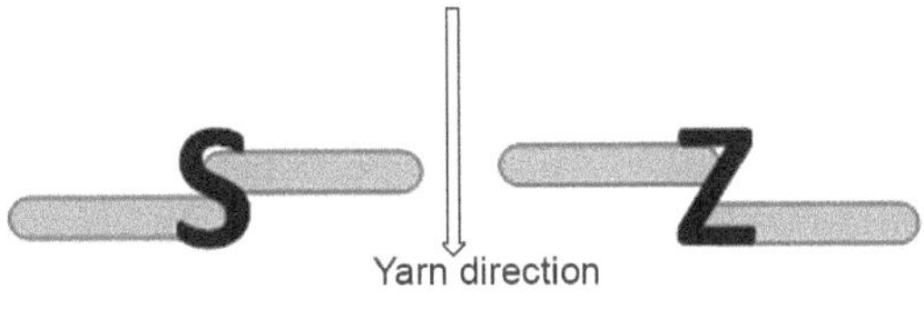

Fig. 2.21: Setting direction of twist

Entry and Exit Discs: The main function of the entry and exit discs is to make sure that the yarn is fully in touch with the top and bottom working discs. This way the full twist potential of the working discs can be exploited. Usually, the discs are highly polished, which means minimum torque and twist loss. Frequently, the exit disc has a distinctive edge. It was developed to counter tight spots in the yarn. When PU discs are being used, the use of ceramic entry and exit discs is nearly mandatory. The ceramic discs protect the soft PU discs against operator errors during the string up of the yarn.

Yarn Path through the Friction Unit: All commercial friction disc units consist of three shafts on which the overlapping discs are being stacked. The yarn travels in a three-dimensional helix from disc to disc. The number of working discs is determined by the processing parameters, such as denier, yarn speed and twist level. A typical set-up for 150/48 denier is a 1-7-1 configuration for hard discs and a 1-4-1 configuration for soft discs (1-7-1 means 1 entry disc, 7 working discs, 1 exit disc).

Fig. 2.22: Motorised friction units on a HY2-V texturising machine, below on-line tension measurement devices

D/Y Ratio

The disc surface speed versus the yarn speed (D/Y) is adjusted to a desired T2/T1 tension ratio. T1 is the tension at the entry of the unit and T2 the exit tension. D/Y ratio is one of the critical process parameters. If the D/Y is too low, the T2 tension would be too high and the T1 tension too low. This can lead to an unstable texturising process. Faster turning discs increase the T1 tension slightly but markedly lower the T2 tension. Processes run successfully even with reversed tension ratios, T2/T1.

Disc Design

In disc designing, special attention is given to the curvature of the working surface. There are two profiles available, the round (C-profile) and the D-profile. The round profile consists of one singe radius, for example, 6.5 mm for a 9-mm-thick disc. The D-profile has two individual radii, for example, 3.5 mm for the edge radius and 9.5 mm for the radius on the working surface. The round profile is used more often, although the D-profile has shown to also give good results.

Fig. 2.23: Profile of discs

Yarn Transport through the Friction Unit

To understand the twisting and feeding action of a friction unit, the basic design of the unit must be understood which will insert an optimum amount of twist. At the same time, the unit will also positively push the yarn through the unit. This twist and push is achieved by having the yarn run at an angle over the surface. The following drawing will help to better understand this action:

Fig. 2.24: arrangement of friction discs

The required numbers of discs are stacked like this together. The discs overlap and the yarn will run along the surface of the discs on the periphery of the highlighted area in the center. The vertical distance is 0.5 mm between the adjoining discs. Important for the twist insertion is the angle of the yarn path over the disc. The steeper the angle, the more twist will be inserted. The smaller the angle, the more the yarn will be pushed along the surface of the disc, which effectively reduces the yarn tension below the friction unit. By choosing the correct angle, a compromise is found, where the yarn is twisted sufficiently for good texture and, due to the pushing action, at relatively low yarn tensions.

The industry optimises processes by selecting the number of discs and in rare cases the diameter and hardness of the discs. Very slightly increased disc diameters will lead to a smaller contact angle of the yarn, thus a lower twist level coupled with low post spindle tension. In some cases, this technique is used to reach higher process speeds.

Figure 2.24 is a picture of a yarn running at 1000 m/min S-twist on a friction unit, which is stacked 1-4-1. The four working discs and the ceramic entry and exit discs can be clearly identified. The yarn direction is from the top to the bottom. The yarn enters, nearly perpendicular, at the ceramic entry disc. From there it moves in a three-dimensional fashion over the polyurethane working discs. The path over the third working disc from the top is visible. The yarn runs at a certain angle over the disc. In reality, this is a very dynamic process, where there cannot be a straight line; the yarn curves around the disc. At the top of the disc the angle is lower, pushing the yarn more, which reduces the tension below. The maximum amount of twist is inserted in the area, where the angle is steeper and the contact is closer to perpendicular.

The D/Y ratio is normally around 1.7–1.8. That means that the disc surface speed is 1.7–1.8 times faster than the linear yarn speed. This speed is a fair amount higher than the rolling speed of the highly twisted yarn bundle. Thus, the yarn is constantly driven by faster moving surface. The wear of PU discs can be attributed to this fact. The same can be said for ceramic, where the fast moving and extremely hard ceramic surface constantly abrades the much softer yarn. This leaves a fine powder, called snow, on the machine.

The bottom disc in this picture is a so-called knife-edge disc, which can be found on many texturising machines. It supposedly helps to eliminate tight spots. It has a polished edge with a very small radius. This disc can be made of either of a very special wear resistant metal or of ceramic.

Crossed Belt Twisting

Crossed belt twisting is another method of friction twisting, also known as nip texturising. The yarn is held at the nip point of two belts mounted at an angle and moving in the opposite directions. These are referred to as nip twisters. Nip twisters were introduced in 1979. The rotation of the belts imparts both twist and a forwarding action to the yarn. The twist insertion rate is controlled by three factors: the angle of the belts to each other, the contact pressure applied to the belts and the V/R, which is the equivalent of the D/Y ratio for stacked discs. It is possible to run two yarns with opposite twist through a single unit and combine them at wind-up.

Heater Arrangements

The heaters are used for heat-setting the crimped filaments. One may adapt a single set of heating elements or two sets where low stretch is required. Heating of the thermoplastic filaments brings them to a molten state while they are in a twisted configuration. This allows the helical angle of twist to be set into the yarn. The heaters may be contact or non-contact in operation. The cooling zone, made up of cooling plates, has two major functions: it allows the yarn bundle to cool, while still in a highly twisted state and it gives stability to the highly twisted yarn bundle between the exit from the primary heater and the entry into the friction unit. This is very important because, if the cooling plates were not present, the yarn would be very unstable in this highly twisted state and have a tendency to balloon resulting in a high break rate.

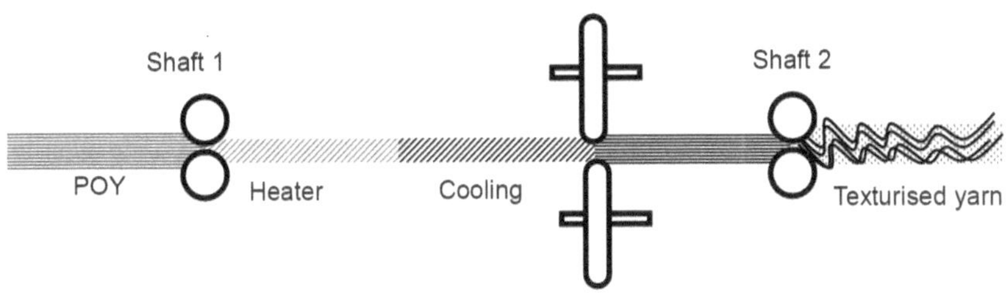

Fig. 2.25: Heater arrangement

In a single-heater friction-type false-twist texturising machine, only one heater is provided.

POY is the standard feed yarn. Shaft 1 is the input point and shaft 2 is the delivery point of the texturising zone. When POY is fed to the machine, the yarn has to be drawn. The speed of shaft 2 is always higher than shaft 1 by the factor of necessary DR for the particular yarn and process. The yarn is simultaneously twisted and drawn. The twisting is done by a friction device such as a rotating set of friction discs. But there are other twisting devices such as belts. After shaft 1, there is a yarn heater, which heats the yarn to a temperature where it can be thermo-set. Right after the heater, a cooling plate is provided which cools the yarn to a substantially lower temperature to permanently thermo-set the twist. As the yarn is released from shaft 2, we observe a voluminous bulked stretch yarn.

Normally, feed yarn is fed into nip rollers which then pass through the single heater. After this, the yarn filaments are cooled before entering the friction discs. After exiting the discs, the yarn passes through another nip roller and is then collected onto a yarn package. This is a close-up of the four feed yarns joining together at the first nip roll system and then passing through the heater. Three stacks of discs provide the twisting of the yarn bundle at a very high rate. The twist inserted flows backwards from the discs through the heater and stops at the nip roll system. Below the friction discs, the yarn twist is removed, and the yarn takes on the crimp appearance. It is then delivered from the machine and formed into a large package.

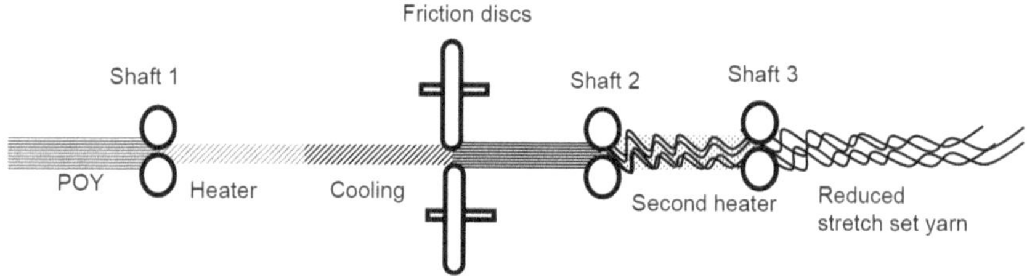

Fig. 2.26: Double Heater Set Yarn Process

In a two-heater machine, after the yarn passes through the twisting device, it can pass through a secondary heater tube usually between 1.0 and 1.3 m in length. At this point, the yarn is heated under controlled relaxation. In the figure, shaft 1 is the input device where POY is fed. Up to shaft 2 the process is similar to single-heater stretching. The highly stretchable yarn is relaxed between shafts 2 and 3 to allow the filaments to crimp slightly. In this condition, the yarn is thermoset once more. This process reduces effectively the undesirable high stretch. The yarn is released from shaft 3. The yarns thus produced have a low crimped high bulk characteristic.

The purpose of the secondary heater is to reduce the amount of skein shrinkage or crimp and stretch left in the yarn after exiting the twist-insertion device. If this is not done, and a secondary heater is not used, it is known as either a single-heater or crimp (high-extension) yarn. Though suitable for certain end uses in this condition, particularly where a fabric of dense construction or stretch is required, it is not suitable for many other applications. The higher the temperature in the secondary heater, the lower the amount of crimp left in the yarn. These are called the double-heater or set yarns. Normally, the temperature of this heater is less than the primary heater. These yarns find use in all types of fabrics. Because of their low shrink character, these yarns give a crisper feel to the fabric and drape better than single-heater fabrics.

Yarn Lubrication

Upon leaving the main frame of the texturising machine, the yarn will be lubricated with antistatic coning oil. This allows the yarn to be processed more efficiently during twisting, knitting or weaving. This lubrication reduces yarn-to-metal friction and yarn-to-yarn friction which helps to maintain proper yarn strength and abrasion resistance during fabric processing. Oil on the yarn also helps provide bundle cohesion, static dissipation and other processing aids. Proper level of lubricant is extremely important for the processing performance of a yarn. The amount of oil applied to the yarn typically varies from 1% to 3%. No coning oil is applied to the yarn if it is going to be dyed in the yarn state.

Package Build-Up

All take-up systems consist of a package support, a drive for the package and a traversing system to prevent the formation of pattern bands in the wound package. Package build must allow for smooth unwinding in subsequent processing. Package density must be controlled, whether the yarn is dyed requiring lower density or when the yarn is used directly onto a high-speed weaving or knitting machine, in which case it would require a higher density. The winding angle chosen for a given package of yarn will affect the way the yarn packs on the package and thus affect dyeing and unwinding efficiency of the yarn.

3

Process Variables and Defects

3.1 Defects

The defects are the undesired effects of a product from the point of view of the customer. If a customer likes a feature in a product, it is termed as a salient feature or effect and if a customer dislikes it, it becomes a defect. Hence, while identifying a defect we need to see it from the perspective of a customer.

The customers give specifications such as basic denier, number of filaments, crimp properties, nodes per unit length, minimum strength required, minimum elasticity and package dimensions. Anything deviating from the specification shall be rejected as poor quality. These parameters can be set precisely. Apart from the specified requirements, there are a number of unspecified but needed requirements to get good results. These are applicable as norms for all texturised yarns. Normally, the complaints received are less in the specified specification area but are more in the unspecified area. As and when a customer experiences a problem, he shall make a complaint. If no problem is raised he shall not complaint. The unspecified requirements of a customer are much bigger compared with the specified requirements.

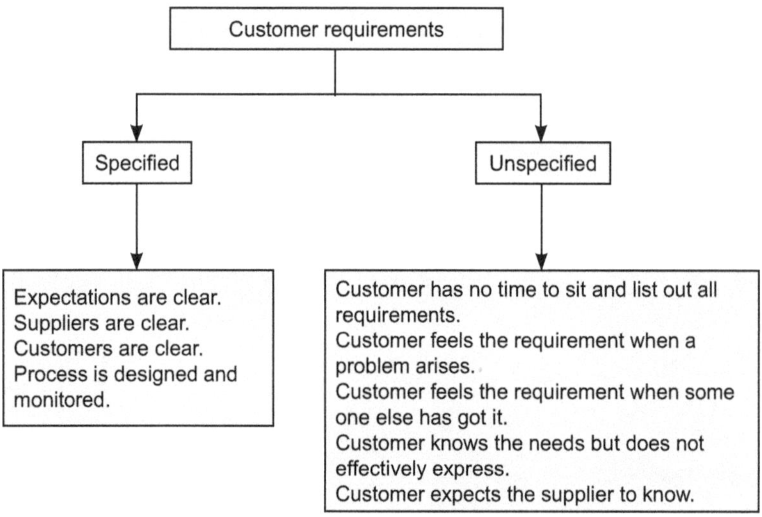

Fig. 3.1: Specified and unspecified requirements of a customer

A defect can arise due to malfunctioning of any machine part, a wrong selection of raw material, wrong selection of process parameters or lapse in following the standard operating procedures. They can be prevented and/or controlled by supervision and checking and strictly adhering to the established standard operating practices while manufacturing. Some customers mention them as conditions in their specifications so that the producer can take required precautions. These requirements are referenced by individual product specifications or procedures, which may also note exceptions and physical properties. The following are some of the examples as given by some customers in the market in the language used by them:

1. *Dirt/grease* – No soiling or grease spots allowed. It is acceptable if the spots can be cleaned off. For dirt on the outside surface air, strip the yarn and make it as waste. For dirt on the ends, clean with sprayer. If dirt does not come off, reject to off grade.
2. *Wound in waste* – None is allowed. Strip if possible to correct. Otherwise reject to rewind.
3. *Damaged/bumped* – None is allowed. Strip if possible to correct. Otherwise reject to rewind.
4. *F oil (coning oil) contamination* – Yarn should not be contaminated with F oil when viewed under a packing table black light. Unless very slight (not immediately visible) strip to clean if possible. Otherwise reject to off-grade.
5. *Broken filaments* – 8 maximum allowable for standard package; 10 maximum allowed for large package.
6. *Texture colour/appearance* – No overly shiny or dull yarn allowed. Dull, bright, dye colour, and so on per limit samples.
7. *Fluorescent oil* – If applicable, there needs to be even coverage when viewed under ultra violet light.
8. *Crossed ends* – Nose end crosses are allowed, unless they appear matted or too numerous to count. Up to two 1 inch crosses on the tail end are allowed or crosses less than ¼ inch from the tube are allowed.
9. *Slubs/loops/kinks* – None is allowed. Reject the materials found with these defects.
10. *Proper wind* – No patterns or bands. No high or falling off edges. No excessive hard/soft packages.
11. *Tube defects* – No crushed, nicked or cut tubes, especially on the nose end.
12. *Ridges/grooves* – No ridges or grooves greater than 1/8 inch high or deep.

13. *Texture* – Yarn should bulk up when stretched several times and relaxed. No end should lack any bulk (raw end). 4 ply through 6 ply may have slight lean ends (end lacks full bulk), otherwise lean ends are not allowed. Also check the tail.
14. *Air tacks* (entanglement) – Minimum 40 tacks/m for regular air. Low air should have low tacks, also depends on specifications.
15. *Twist* – For single-ply yarns only, Z-twist must rotate clockwise when allowed to relax and S-twist rotates counter-clockwise.
16. *Proper ply* – Count the number of ends if the yarn is three ply or more. Strip the yarn using compressed air to correct if possible. Also check the tail.
17. *Latching* – Filaments or plies that separate when winding off package are not allowed.
18. *No-tail* – If there is no tail, pack one large no-tail package per layer. If excess no-tail packages are found then reject to rewind. The minimum tail length is one wrap around the tube. If the tail is too short, take a short length of the yarn and tie it on to the tail. For parallel yarn, small and no-tails are allowed.
19. *Multiple tails* – If there are multiple tails, then correct if possible. Otherwise reject to rewind.
20. *Tube clearance* from yarn roll to tail end of tube should be between 10 and 15 cm maximum (5/8 inch nominal target.)
21. *Oversize or small package* – Check suspect packages with appropriate gauge, scale, diameter tape or balance. Do not put excessive minimum size packages in one case. See specific product specification for other special sizes and weights. Allow one small layer.
22. *Packaging* – Yarn packages are bagged with high-density vented poly bags and packed into box, tail end down, so there is as little movement as possible. The packages are not forced into the box. The box goes on conveyor to scale open with flaps taped down and inspector card placed in one of the tubes on the top.
23. *Weighing* – Yarn is weighed as per agreed procedure (with or without allowance for moisture regain or finish). The case is weighed to determine gross weight. The net weight is calculated based on the number of the tubes and standard tare weight of all the packaging materials.
24. *Case labeling* – Cases are labeled and individually numbered with computer transfer printed, stuck on three places. In addition, each case has tube label and shipping label. The labels have an important

warning; 'If the Lot or Merge number on the container differs from previous shipments, use separate unless your test prove this unnecessary'.

3.2 Understanding the Defects and Their Causes

The raw materials used play a very important role in getting the required quality of output product. Faults in spinning process such as the changes in polymer properties, spinning temperature, quench air conditions or take-up speed cause problems in downstream processes and defects in the end product. This can be controlled by POY producer.

While trying to understand the defects in texturised yarns, it is necessary to understand in what way it affects the customer or the next process. The requirements of the yarn are grouped into four main categories by J. W. S. Hearle, L. Hollick and D. K. Wilson in their book 'Yarn Texturising Technology'; they are as follows:

(a) Those that affect the tensile strength
(b) Those that affect the behaviour in the fabric
(c) Those that affect the dye uniformity
(d) Those that affect the performance during knitting, warping, weaving and so on.

They observe that no one attribute of the texturised yarn can stand alone. Each property of a yarn has an impact on another. Hence, all the properties are to be studied. The normal tests conducted to find the suitability of a texturised yarn are tensile testing, denier testing, yarn skein shrinkage, degree of bulk, degree of crimp, degree of texture, stability of bulk, stability of crimp, stability of texture, assessment of expected cover in fabric, package to package dye uniformity and visual inspection before despatch.

Normally, an inspector in the visual inspection observes for

(a) Dirt marks on the yarn
(b) Secure transfer tail
(c) Package build faults that would lead to unwinding problems
(d) Presence of high number of broken filaments
(e) Uniform intermingling, if present
(f) Damaged tube, container or bobbin
(g) Package size and weight
(h) Correct tube colour for the product
(i) Lot number mentioned and the actual
(j) Markings made on cartons

Fig. 3.2: Package dimension variations

Fig. 3.3: Texturised yarns kept for visual inspection

Fig. 3.4: Non-standard colour codification

Guarantee/disclaimers, if any, specific to the material and the process involved are verified and stated clearly so that there shall not be any dispute.

3.3 Factors Influencing the Properties of Texturised Yarns

There are three main factors on which the various properties of the texturised yarn depend. They are machine parameters, parent yarn conditions and process parameters.

Machine Parameters: In a texturising machine, various machine parameters such as the type of heater, length of heater, form of heat transmission, feed system, type of false-twisting mechanism used, type of spindle, type of friction discs and package building mechanism influence the properties of the texturised yarn.

Parent Yarn Conditions: Parent yarn conditions such as the raw material, denier, lustre, number of filaments, extent of orientation in POY, spin-finish and cross section influence the property of the texturised yarn. While using LOY, it is essential to verify the shelf-life. Older the yarn, it is difficult to texture. LOY is applicable to only composite units having melt spinning and draw texturising.

Process Parameters: Process parameters, which are variable and can be set precisely considering the material in use and the required effects, have an influence on the properties of the texturised yarns obtained. The process variables such as the speed of production, speed of twisting, temperature of heat-setting, time of dwell for the material in heat-setting area, cooling temperature, draw tension, tension in the heater zone, direction of twist, twist level and rate of feeding are responsible for changing the yarn properties. If they are not set properly, the entire yarn shall be considered as defective yarn. Let us have a look into some of the process variables.

3.2.1 False-Twist Process Variables

1. **DR** – The DR is the amount of yarn stretched between the input roller and the output roller which follows the twisting mechanism. The metres of yarn traveling per minute through the output roller divided by the metres of yarn traveling per minute through the input roller is how the DR is calculated. The DR will determine the final texturised yarn elongation, the yarn denier and the yarn tenacity which is a value calculated from the yarn breaking strength and the resultant denier. Higher DRs will create excessive broken filaments, low dye uptake

and reduction in residual shrinkage. Low DRs create the opposite of these.

2. **Primary heater temperature** – Increasing primary heater temperature will increase yarn bulk and filament breaks while decreasing dye uptake. Decreasing the temperature will create the opposite of these factors. It depends on the yarn properties that are required.
3. **D/Y ratio** – The D/Y ratio is the ratio of speeds between the friction discs and the linear speed of the yarn. The most common method of changing the amount of twist inserted in the yarn is by changing the speed of the discs, keeping the number of discs and the spacing between them constant.
4. **Second heater temperature and overfeed** – The higher the temperature in the second heater, the lower will be the final yarn shrinkage. If a high second heater overfeed is employed, the tension on the yarn in the heater tube is low allowing the heat available to have its maximum effect.
5. **Package build** – Off-winding performance of the yarn is crucial to meeting customer needs. Factors affecting this performance and the form of a package include yarn skein shrinkage, yarn denier, intermingling level, take-up overfeed, wind angle, taper angle, traverse stroke length and winding tension.

3.2.2 Air Jet Process Variables

1. **DR:** The DR is the amount the yarn stretched before entering the texturising zone. When feeding two yarns, the DR has to be set for each yarn. In addition to overfeed and subsequent stretching and relaxing, the DR determines the final yarn denier. Over-drawing causes broken filaments while under-drawing causes the yarn to be plastic or non-elastic when stretched under load. This may be desirable when a fabric is to be stretched into the shape of a seat or some other object.
2. **Hot-pin or Draw-pin:** Polyester is widely available in the form of POY and, therefore, any texturising process suitable for this market has to include the means for drawing POY. Common practice is to use a heated draw-pin located between the first two draw rolls which provides a fixed draw point and a means of heating the POY. Typical temperature for polyester ranges from 135°C to 160°C. This higher temperature reduces residual shrinkage.

3. **Yarn overfeed**: Yarn overfeed has a direct influence on the bulk or specific volume of the yarn. Different yarns can have different overfeed rates and create core-effect type of the yarns. Surface loops can be formed to make spun-like yarns. Changing the core yarn overfeed has a direct influence on the strength of the texturised yarn as well as its stability or resistance to removal of the surface loop structure under load. Because the core overfeed affects the length of the texturised yarn, it has a major influence on the denier of the yarn. The yarn overfeed determines the size of the loops. It also has an indirect influence on the number of loops per unit length and also on the denier.
4. **Yarn wetting**: Heavier yarn requires more water to be applied. Water helps to reduce the yarn-to-yarn friction with the feeding of multiple yarns under different tensions and feeding rates. Water and spray are collected in the jet box and removed by means of a gravity drainage system.
5. **Air-texturising jet**: From 500 to 5000 denier yarns can be used with either an axial or radial type of jet. Radial jets are used with lower overfeeds and higher processing speeds. Axial jets operate with considerably higher net overfeeds in the range of 4%–20% to impart stability to the texturised yarn. A final texturised denier of 1215 prepared for weaving could be made from feeder yarn of 750 deniers at the point of entry into the jet. This means that the total increase in denier is 60%.
6. **Mechanical stabilisation:** The stability of an air-jet texturised yarn is improved considerably by stretching after the texturising zone. This stretching is done by using at least two yarn feed units after the jet box. The higher the (core) overfeed, the greater will be the effect of stretching on the yarn stability. From 2% to 10% is normal. Heat may be applied within the mechanical stabilisation zone as a means of inducing tension in the yarn.
7. **Heat-setting:** Air-jet texturised yarns are much denser than false-twist texturised yarns of the same denier and, therefore, the effect of heat-setting is not as noticeable in the end product. Nevertheless, the effect of applying heat means that overfeeds of up to 10% is used, depending on the residual shrinkage of the texturised yarn. The applied overfeed must be absorbed by the shrinkage of the yarn or else the process would break down.
8. **Yarn build and lubrication:** Due to air-jet texturised yarn having a much lower stretch or elongation, the take-up overfeeds applied

are lower. Methods of applying lubricant to the yarn and varying the quantity are identical to those for the false-twist texturised yarns.

9. **Air-jet-texturised yarn properties:** Air-jet-texturised yarn properties such as instability, physical bulk, hot water shrinkage, core diameter, loop size, loop frequency, tenacity, breaking extension and initial modulus are affected by process parameters such as overfeed, air pressure and heater temperature. Overfeed has much greater effect than other process variables. The texturised yarns produced from parent yarns possessing small amounts of pre-twist have a high instability because of the lack of binding twist for securing the entangled structure. Increasing overfeed greatly increases the number and size of loops, and this causes greater instability. An increase in the applied air pressure also increases the yarn instability.
10. **Initial filament twist and overfeed:** The initial filament twist in the parent yarn redistributes itself as snarled loops and the overfed yarn contributes to the size of such loops. The deflexion of the filaments by air-drag forces also makes a contribution to a looping effect. Increasing the yarn overfeed increases both the loop size and the loop frequency. Increasing the yarn pre-twist decreases the loop size and increases the loop frequency. Increasing the air pressure also decreases the loop size and increases the loop frequency. Overfeed is the greatest single contributor to physical bulk, because both loop size and loop frequency increase with overfeed, occluded air spaces being thereby created in yarn and fabric form; conversely, with varying twists and air pressures, the effects of decreasing loop size and increasing loop frequency are to some extent self-balancing in their contribution to physical bulk.
11. **Tenacity of yarn:** The tenacity of the air-texturised yarns tends to decrease for the increasing values of all the main processing variables considered. Increasing the yarn pre-twist tends to reduce the percentage elongation at break, but independent increases in the supply yarn overfeed and the applied air pressure both tend to increase this characteristic.

3.4 Normal Concerns in the Texturised Yarns

Some of the normal concerns in texturised yarns are as follows:

- **Yarn breaks** – The yarn break rate is the number of yarn breaks that take place per unit weight of the texturised yarn produced. Yarn breaks can create short yarn packages, lost production and increased labour and packaging costs. The breaks caused by low-quality POY yarn should be segregated and machine speeds should be lowered. The process-related causes can be an excessively high DR, excessively high primary heater temperature and high processing speeds, causing instability and surging in the thread path. Poor machine maintenance creating things such as worn or damaged ceramic discs, poor thread path alignment and damaged nip rolls and aprons can also cause yarn breaks.

- **Intermingling faults** – Intermingling faults mainly involve the frequency or strength of the intermingling knot and the irregularity or unevenness of intermingling along the length of the texturised yarn. The choice of jet must match with the overall denier, filament denier and the speed of the process for the intermingling to be strong and uniform. Process conditions such as air pressure, yarn tension, speed and jet location should be optimised. Age and cleanliness of the jets must also be considered. Clogging of air jets also give uneven intermingling.
- **Bulk variation** – Bulk variation can be either end-to-end variation around the texturising machine or variability of the yarn along the yarn length. The causes can be uneven heat penetration due to the yarn core being denser than the surface filaments. This is more likely when low filament denier yarn is being texturised. Other causes of bulk variation may be low DR or incorrect rate of twist insertion for the particular process. This may be remedied by changing either the D/Y ratio or the number of friction discs employed.
- **Dye variation** – It may be an overall drift in shade towards light or dark dye or an increase in end-to-end variation within the machine. The change in polymer, POY variations, dirty primary heaters, wear of polyurethane discs, yarn not fully drawn and heater temperature variation can all lead to dye variation.
- **Package density problems** – This problem is particularly a concern for packages designated for dyeing. Yarn shrinkage and intermingling properties must be correct. Incorrect winding tension, errors in the thread path, incorrect taper angle and improper wind angle tension variation can all lead to wrong package density for a given product of the texturised yarn. These factors must meet the required specifications.
- **Surging** – Instability in the thread line is termed as surging. It manifests itself in a lean untexturised appearance. Surging can be caused by properties of the POY feed yarn, incorrect machine parameters, DR

being too low, wrong throughput speed for a given product, irregular application of spin-finish and irregular yarn orientation along the length of the yarn.

- **Package build faults** – One of the more common package build faults is bulging, caused by using small wind angles. Another is webbing or scrambling of the yarns on the package, many times caused by a too low wind angle for intermingled products. High-density packages produced with high take-up tensions will also result in webbing. Overthrown ends, ridges, shouldering and no-tail packages are other common package build faults.
- **Broken filaments** – Texturised broken filaments may be related to the yarn type, POY contributing factors, mechanical damage within the thread path or the settings of the texturising machine on which the yarn is produced. Broken filaments are a more common occurrence on low denier rather than high denier products.
- **Tight spots** – Take the form of a very short length of the untexturised yarn and are most easily detected by inspecting long lengths of yarn, examining a knitted sleeve or by on-line monitoring. The causes of tight spots are those mentioned previously for surging and the same remedial actions can be taken.
- **Instability of ATYs** – If the loops of ATYs are pulled out during further processing, the yarn bulk will be reduced, and if this bulk reduction takes place selectively in certain sections of the yarn, the irregularity of the product will be increased.

3.5 Specific Performance Properties of Texturised Yarns

The following are the specific performance properties of a texturised yarn; whether it is specified by the customer or not, it is the duty of the yarn manufacturer to maintain and monitor.

- **Tack retention**: The ability of tacks to withstand stretching. This is important when processing imparts different degrees of strain on the yarn. Tack retention quantifies the 'strength' of the tack to withstand strain and, therefore, provide bundle cohesion for processing'.
- **Twist direction**: The rotational direction by which twist is present in a yarn. It can be designated as 'S' or 'Z'.
- **Twist level:** The number of turns of twist in a yarn per a given length of yarn. For the twisted yarns, the level of twist is critical for the

degree of bundle cohesion needed for processing and yarn strength. In the texturising process, the twist level has a significant impact on the crimp level of the yarn.

- **Torque:** The tendency of a yarn to rotate and kink upon itself in an effort to relieve twist. The impact of torque, causing kinks in the yarn, can lead to processing problems. Torque can often be observed in the fabrics. Torque can be balanced by plying 'S' and 'Z' yarns, or minimised with jets.
- **Skein shrinkage:** The length contraction of a skein of yarn held at a prescribed load and after heating under prescribed conditions. Tests are designed to simulate various down-stream processing conditions. Some tests simulate the restraining forces experienced in fabric dyeing and finishing, and is used to design texturised yarns for different types of fabric applications.
- **Residual skein shrinkage:** The residual shrinkage test utilises a much lower restraining load than other tests and is a better indicator of the ultimate shrinkage potential of the yarn.
- **Differences in reflectance:** Reflectance differences lead to barre in the fabrics. E. R. Cairns, H. A. Davis and J. W. Coryell observed that double-knit barre was caused by texturised yarn reflectance differences in the knit structure. Depending on the detailed arrangement of these differences, barre is seen as continuous or random and either as single-end streaks or bands. Based on the studies carried out with a research grade spectrophotometer, they also found that yarn reflectance differences are caused by the differences in key texturised yarn properties such as bulk, cross section and loop size.

The most important quality problems of BCF yarns are the differences in crimp and shrinkage properties leading to unevenness of the carpet both with regard to yarn bulk and dyeability (for polyamide yarns).

4
Normal Defects and Their Causes

In this chapter, some of the normally observed defects in false-twist texturising are discussed.

The defects normally found in a texturised yarn can be grouped as follows:

(a) Yarn defects due to process variation
(b) Yarn defects due to setting variations
(c) Yarn damages
(d) Winding defects due to process variations
(e) Winding defects due to setting
(f) Defects due to damaged parts in machine
(g) Defects due to machine setting
(h) Defects due to human work practices

The following are the grouping done while observing the actual defects in a texturised yarn and process.

Group	Sub-group	Defect/problem
Yarn	Process variation	Untexturised yarn
Yarn	Process variation	Un-drawn yarn
Yarn	Process variation	Bulk variation
Yarn	Process variation	Nip variation
Yarn	Process variation	Tight spot
Yarn	Process variation	Loops
Yarn	Process variation	POY lapping
Yarn	Process variation	Snarling
Yarn	Process variation	Too much spin-finish
Yarn	Process variation	Heater deposits
Yarn	Process variation	Snow formation
Yarn	Setting	Nips per metre improper
Yarn	Setting	Hard nips
Yarn	Setting	Crimp contraction too low
Yarn	Setting	Crimp rigidity too low

Yarn	Setting	Surging
Yarn	Setting	Yarn comes out of input/intermediate/output nip roller
Yarn	Setting	Yarn not passing through the stainless steel tube
Yarn	Setting	Cutter not functioning
Yarn	Setting	Yarn wrapping
Yarn	Yarn damage	Broken filaments
Yarn	Yarn damage	Filmentation
Yarn	Yarn damage	Weak yarn
Yarn	Yarn damage	Yarn breaks
Winding	Process variation	Poor rewind package
Winding	Setting	Jali/slippage of coils in the side
Winding	Setting	Bulging
Winding	Setting	Slippage
Winding	Setting	Tight winding
Winding	Setting	Bad build-up
Winding	Setting	Stroke length disturbances
Winding	Setting	Shore hardness of the package is not onto the standard
Machine	Damaged parts	Damaged disc
Machine	Damaged parts	Damaged thread guides
Machine	Damaged parts	Damaged disc driving (Positorq) belt
Machine	Damaged parts	Destruction of gear belts
Machine	Damaged parts	Bearing damage at friction disc unit
Machine	Damaged parts	Excessive vibration/noise from the feed shaft
Machine	Damaged parts	Hairy yarn
Machine	Setting	Speed variation at friction disc unit
Machine	Setting	Top/bottom heater guides not moving with detent mechanism
Machine	Setting	Positorq unit stops
Machine	Setting	Slippage between Positorq wharve and drive belt
Machine	Setting	Noisy Positorq
Human	Work practice	Dirty or oily package
Human	Work practice	Poor storage

Human	Work practice	Wrong threading
Human	Work practice	Damaged package
Human	Work practice	Poor package unwinding
Human	Work practice	Oil spot
Human	Work practice	Without oil

The description of the defects observed are the possible reasons of the defects, the consequences of defects or the effects of the defects, and the remedial actions to be taken to avoid those defects are explained as follows.

4.1 Yarn Defects due to Process Variation

4.1.1 Untexturised Yarn

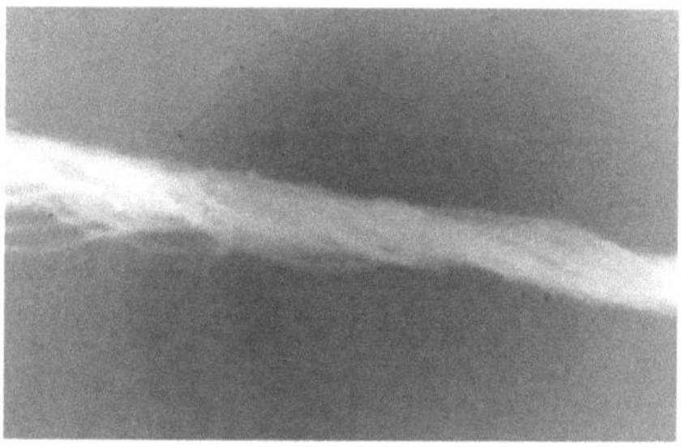

Fig. 4.1.1: Untexturised Yarn

Fig. 4.1.2: Knitted and dyed fabric of Untexturised yarn

Description

- Long length of the texturised yarn having little or no bulk: when knitted and dyed, we get short pattern effect which is permanent.
- Missing of texturised effect for long lengths of a texturised yarn.

Causes

- ***High U% of POY:*** Higher unevenness of POY results in instability in the yarn while texturising and leads to tension variation. Because of this tension variation, the yarn momentarily loses contact with the friction disc leading to the untexturised yarn. The length of the untexturised portion in the yarn depends on the magnitude of instability.
- ***High spin-finish oil% in POY***: The spin-finish should evaporate in the primary heater and the yarn should contain bare minimum percent of spin-finish before approaching the twisting assembly in draw texturising. Higher content of spin-finish in the yarn while approaching the friction twisting units, the oil causes rotational slip of the yarn leading to untexturised yarn.
- ***Low D/Y ratio:*** The D/Y ratio indicates the ratio of disc speed to a given yarn speed. When the D/Y ratio is low, the twisting assembly fails to impart the desired twist because of low rotational speed of the friction discs, resulting in untexturised yarn.
- ***Spindle vibration:*** A defective bearing in the twisting assembly causes vibration in the twisting assembly as well as lowering in the rotational speed of the friction discs. This results in lower twist and untexturised portions in the yarn.
- ***Splicing:*** Splicing of POY is a normal practice to improve the machine efficiency and productivity. The diameter of the spliced portion is more compared with that of a single yarn. When this portion comes in contact with the friction discs, the effective torque is increased and also the twist. This is followed by a sudden drop in yarn diameter as well as contact area in the friction discs. This causes a momentary drop in friction torque; the instability continues for an appreciable time span leading to long untexturised yarn.
- ***Failure of heat-setting:*** Momentary failure of power to the heat-setting unit allows it to become cool.

Effects

- When knitted and dyed, it shows as a permanent pattern effect.

Remedies

- POY used should be uniform with U% of less than 1%.
- POY having spin-finish below 0.40% to be used. Uniformity of spin-finish is very important.
- Optimise D/Y ratio.

- Optimise DR.
- Authentic and prescribed bearing should be used. Do not change the bearing supplier or manufacturer.
- Each and every position should be thoroughly checked during maintenance schedule.
- POY weight should be adjusted to the multiple of texturised yarn bobbin weight.
- If at all splicing is required, the overlapped portion should be kept minimum possible.

4.1.2 Undrawn Yarn

Fig. 4.1.2: Knitted and dyed fabric of undrawn yarn

Description

- The texturised yarn which escapes the effects of drawing is called the undrawn yarn.
- The undrawn/partially drawn gives high shrinkage and low tenacity causes defective fabric.

Causes

- ***Yarn not in the grip of the first nip roller:*** When the throw (movement) of the traverse mechanism fitted at the first nip roller region has larger amplitude, the yarn often comes out from the grip of the first roller causing no drawing of the yarn leading to UDY.
- ***Weak spring of intermediate nip roller:*** Weak spring in the intermediate nip roller prevents it from proper loading and creates gap in dynamic conditions resulting in poorly drawn or UDY.

Effects

- When knitted and dyed, UDY produces pattern effects.

Remedies

- Adjust traverse mechanism and ensure that the yarn does not come to the edge of the rollers.
- Periodically check the pressure on the nip rollers and ensure it as adequate.
- Change the spring of the intermediate nip roller if found weak.

4.1.3 Bulk Variation

Fig. 4.1.3: Bulk variation

Description

- Differential bulk along the length of the yarn is known as bulk variation.

Causes

- ***Low DR:*** The yarn property if it is not homogeneous along the yarn length is mainly responsible for thread line tension variation at high-speed texturising. At low DR, the variation in thread line tension aggravates resulting in differential twist along the yarn length, leading to the variation in yarn bulk.
- ***Tension variations:*** The yarn property is not homogeneous along the yarn length, which is mainly responsible for thread line tension variation at high-speed texturising.

- ***Low DR:*** At low DR, the variation in the thread line tension aggravates resulting in differential twist along the yarn length, leading to the variation in bulk.

Effects

Causes the variation in weight per unit area in the fabric.

Bulk variations give differences in hand and feel of the fabric.

Remedies

Optimised DR to be used.

4.1.4 Nip Variation

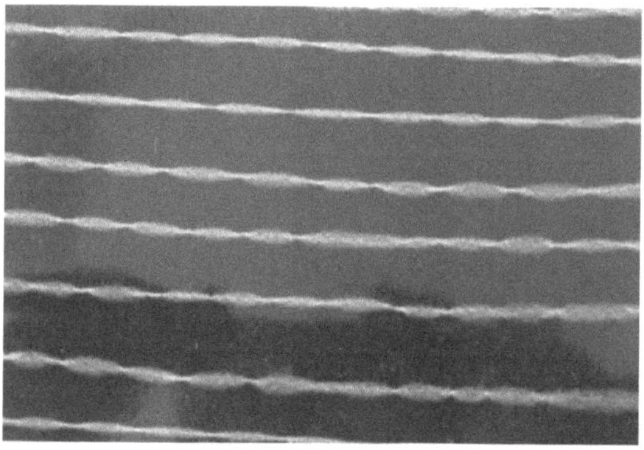

Fig. 4.1.4: Nip variation

Description

- Non-uniformity in the number of nips along the yarn length is known as nip variation.

Causes

- ***Improper jet selection:*** The quality of nips or intermingling depends on the diameter of the nozzle. The higher the yarn denier, the more is the requirement of the compressed air for intermingling and so also the nozzle diameter. The jet having lower nozzle diameter leads to inadequate air supply, thus causing the variation in the number of nips per unit length.
- ***Chocking of nozzle:*** Compressed air sometimes contains contaminants such as grease, oil, along with mud and sand of unclean rusted air conveying pipe. This chokes the nozzle mouth and inhibits the air

flow leading to the variation in the number of pipes along the yarn length.

Effects

- Nip variation causes pattern effect in the fabrics by taking dyes in a different way.

Remedies

- Selection of right type of nozzle.
- Use jet pipe cleaned by organic solvent.
- Periodically clean the intermingling jets in ultrasonic fluid bath.

4.1.5 Tight Spot

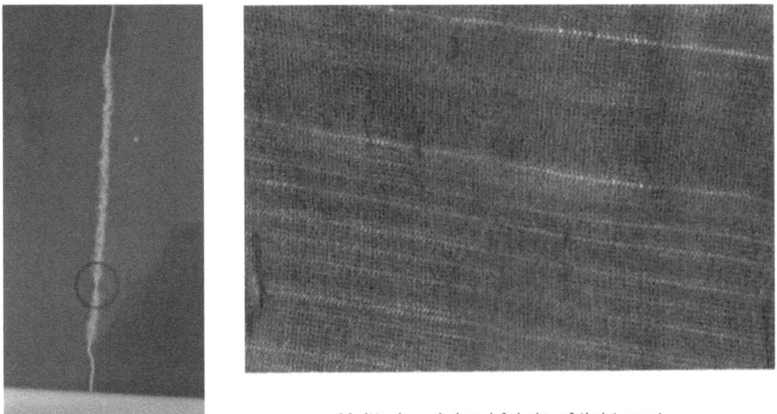

Tight spot

Knitted and dyed fabric of tight spot

Fig. 4.1.5: Tight spot

Description

- Short length of compact yarn caused by the pressure of high twist.

Causes

- ***Low DR:*** Any yarn is not perfectly homogeneous in every respect. At low DR, the thread line tension also becomes low and slightest variation in any of the yarn property causes rotational slippage leading to a tight spot.
- ***High D/Y ratio or D/Y ratio not optimum:*** Rotational slippage, which is caused at high D/Y ratio, develops a momentary instability causing

real twist appearing in the post spindle zone leading to tight spot. On the other hand, at low DR, the vibration of the machine is sufficient to cause thread line instability in the form of thread line tension variation, causing rotational and translational slippage of the yarn at friction twisting unit, in turn responsible for real twist to appear in the post twisting zone thereby leading to the generation of tight spot in the texturised yarn.

- ***Increased threading:*** Increased threading increases tension in the yarn which is not uniform, leading to real twists in the yarn at some spots.
- ***High U% of POY:*** Uneven POY leads to improper flow of twist in the yarn, developing some high-twist thin places and low-twist thick places.
- ***Worn out nip roller:*** Worn out nip rollers give jerks to the yarn resulting in uneven twist distribution leading to some real twist portions in the yarn.
- ***Process parameters unstable***: Unstable process parameters such as tension variations and pressure variations on feed and delivery roller nips result in tight spots.
- ***Filaments sticking together in texturising zone***: Momentary sticking of filament in the texturising zone holds the movement of yarn running at high speed resulting in a tight spot.

Effects

- The spots are seen on the fabric. The tight spot when dyed takes more dye and looks darker.
- Rotational slippage of the yarn is caused at high D/Y ratio.
- Develops a momentary instability causing real twist appearing in the post spindle zone leading to tight spot.
- The yarn is not perfectly homogeneous in every respect. At low DR, the thread line tension also becomes low and slight variation in any of the yarn property causes rotational slippage leading to tight spot.

Remedies

- Check D/Y ratio, D/R ratio and knife setting and optimise D/Y and D/R.
- Reduce the temperature of the texturising heater.
- Check for film build-up on disc surface.
- Check unstable running of the filament.

4.1.6 Loops

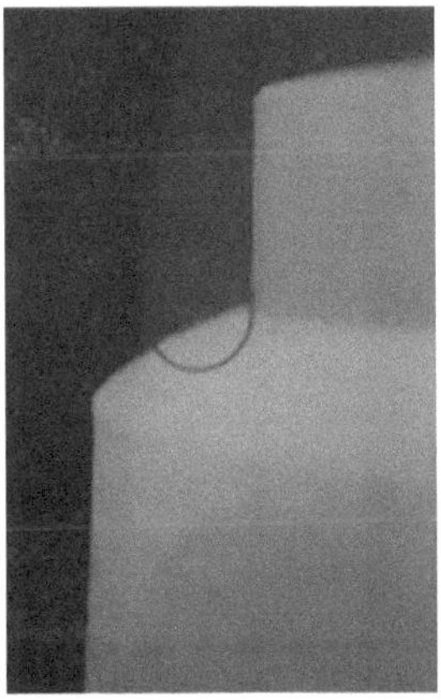

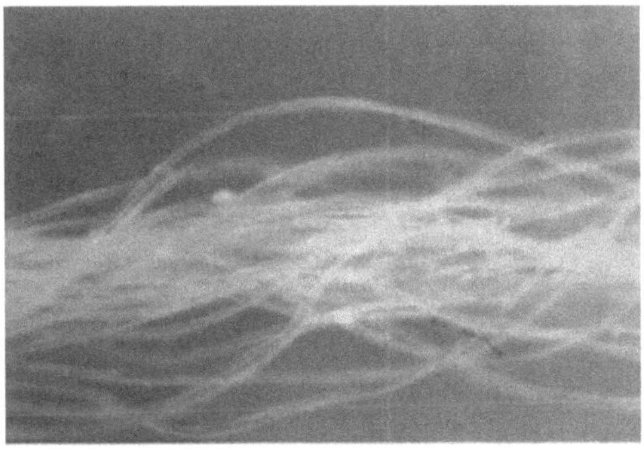

Loops

Fig. 4.1.6: Loops

Description

- Protruding of continuous filament from the yarn as well as from the package.

Causes

- ***Low primary heater temperature:*** Multifilament yarn on its passage through the primary heater experiences differential heat exchange causing differential thermal state of the filaments. This inhomogeneity is aggravated at low primary heater temperature. When a yarn having differential thermal state of the constituent filaments comes in contact with the friction disc, some of the filaments tend to deform in the form of unusual extension and appears as loops.
- ***Low stabilising overfeed:*** In the secondary heater, the mechanical stresses, which were imparted in the yarn during twisting, are released and the yarn simultaneously shrinks. The tension of the yarn plays an important role for uniform heat exchange. The softer tension resulted from low stabilising overfeed; differential shrinkage potential of the yarn is aggravated in loopy yarn.
- Low quality of POY elongation and spin-finish.
- High D/Y ratio.
- ***Yarn tension in setting zone too low***: Irregular and a too low yarn tension makes some filament becoming loose and come to the surface generating a loop.

Effects

- Broken filaments during twisting and breakages on the loom.

Remedies

- Check secondary heater temperature and optimise it.
- Check D/Y ratio.
- Check filament breaks.
- Increase D/R ratio.
- Check guides, thread line and friction disc.
- Check evenness of finish application.
- Increase quality of spin-finish.
- Optimised stabilising overfeed to be used.

4.1.7 POY Lapping

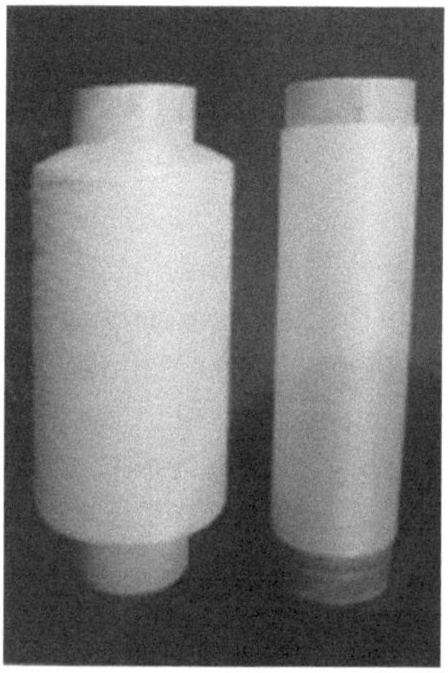

Fig. 4.1.7: POY lapping

Description

- Sticking of filament (POY) to roller surface.

Causes

- POY gets out of input nip roller and gets lapped when broken.
- Traverse not set properly making the POY to go out of nip and then getting lapped on the rollers running at high speed.
- Yarn guide disturbed.

Effects

- It cannot be used as it is not texturised or stretched.

Remedies

- Following correct operating practices.
- Replace the existing guides with new guides.
- Set traverse with the help of reduction gear box.

4.1.8 Snarling

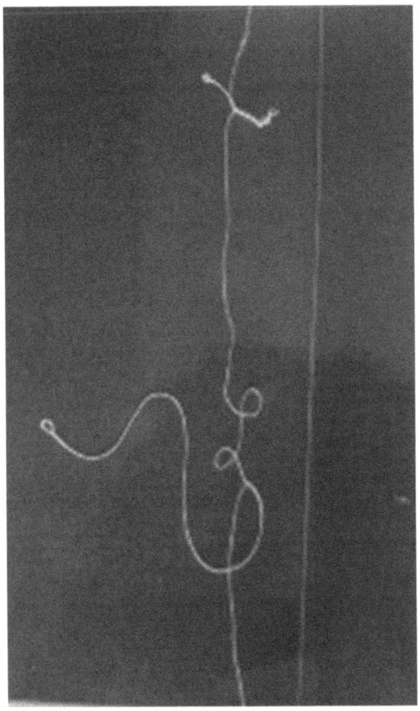

Fig. 4.1.8: Snarling

Description

- Twisting on to itself forming twisted loop.

Causes

- Yarn gets out of nip roller, may be due to improper setting or the guide becoming loose.
- Twisted portion of the yarn is let loose before heat-setting in the heater and also known as surging.

Effects

- It cannot be used as the snarls entangle with adjacent yarns and cause breaks.
- Difficult to process.

Remedies

- Set the yarn properly at the centre of the nip roller.
- Provide a guide or set the existing guide to ensure the yarn does not come out of nip rollers.
- Practicing good work methods and operating instructions.

4.1.9 Too Much Spin-Finish (Coning Oil on the Yarn)

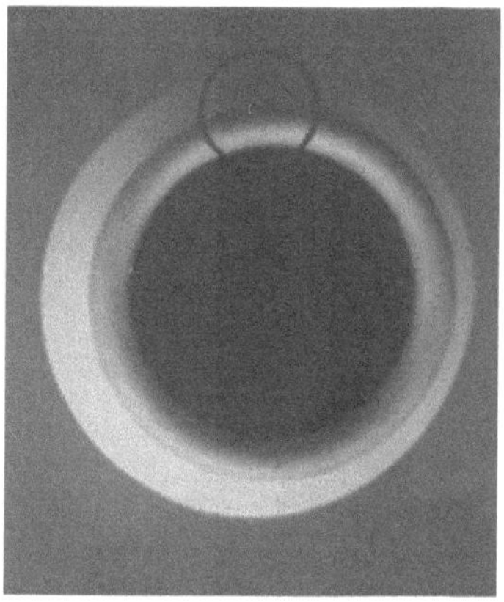

Fig. 4.1.9: Too much spin-finish

Description

- Excessive coning oil on yarn package.

Causes

- Yarn path setting disturbed.
- The oiler roller speed (RPM) increased.
- Yarn waste accumulation on the oil roller.

Effects

- The denier of the texturised yarn increases.
- Splashing of oil during the downstream process of winding and twisting.

Remedies

- Check yarn path setting and correct it.
- Check the oiler roller speed and set it on all positions.

4.1.10 Heater Deposits

Descriptions

- Deposits collected on the heater.

Causes

- The heaters are not cleaned thoroughly while changing the lots.

Effects

- Production down because the yarn will not run.
- Soiling of the heaters due to thermal degradation of spin-finish of the feeder yarn.

Remedies

- The heater should be kept clean.
- Before starting new lot, clean the heaters with the cleaning tools and ensure all degraded oil particles are removed.

4.1.11 Snow Formation

Description

- Collection of some white powdery substances on the parts of the machine which may go along with the texturised yarn.

Causes

- More spin-finish in the POY.
- Insufficient film strength of spin-finish.
- Rough and abrasive friction disc.
- Higher digomers, monomers of polyester yarn too low.
- Unstable spin-finish component in the POY yarn.
- High yarn temperature at friction disc.

Effects

- Soiling of the roller shafts and friction spindles due to snow formation on texturising machine and on the yarn.

Remedies

- Reduce quantity of spin-finish.
- Use suitable spin-finish.
- Use discs with smoother surface.
- Improve cooling of the yarn before friction disc unit.

4.2 Yarn Defects due to Setting Variations

4.2.1 Nips Per Metre Improper

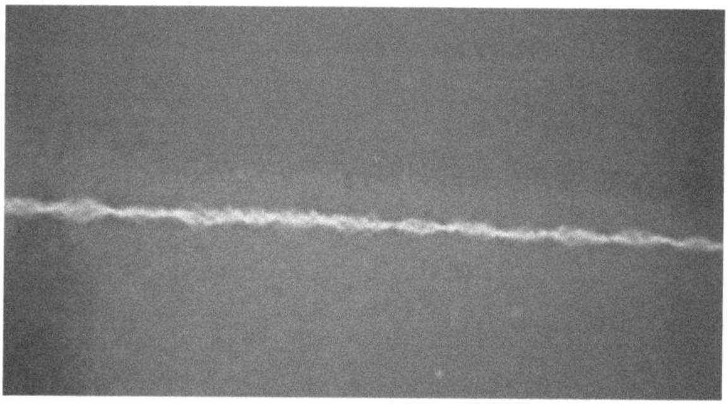

Fig. 4.2.1: Nips per metre not uniform

Description

- Uneven nips formed during texturising and intermingling.

Causes

- Air pressure is not consistent.
- Jet problem – jet is not smooth and the variation in jet pressure due to clogged air passage through the air jets.
- Tension variation.

Effects

- The yarn shall be graded as substandard and fetches a very low price or gets rejected by the customer.

Remedies

- Check air pressure.
- Check condition of jets.
- Check and maintain uniform tension.

4.2.2 Hard Nips

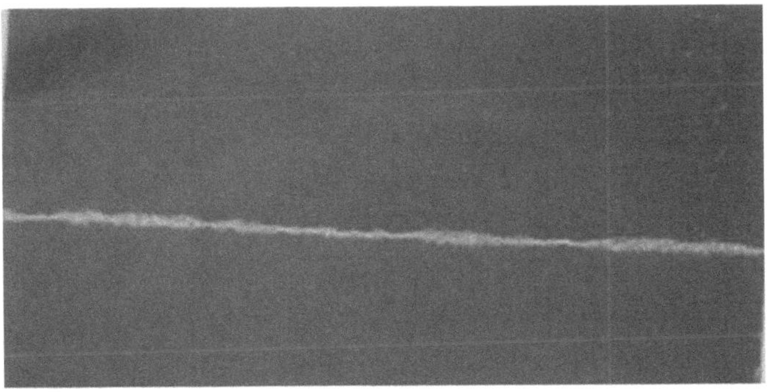

Fig. 4.2.2: Hard nips

Description

- Highly twisted portion of the filament spread irregularly on the yarn.

Causes

- High pressure on nip rollers compressing the yarn.

Effects

- The filaments do not open while texturising either by twist–untwist method or by air jet.
- Dye variation in the fabric.

Remedies

- Check air pressure and adjust.

4.2.3 Crimp Contraction Too Low

Description

- The yarn is having a twist contraction much lower than the specified.

Causes

- Twist insertion in the texturising zone is too low.
- Yarn tension in the setting zone is too high.
- Too high yarn temperature at friction disc.

Effects

- In a fabric, when other yarns are contracting at a certain level and this yarn is not contracting, the fabric shall not have a stable surface resulting in dyeing defects in the fabric. Hence, it is unacceptable to the customer.

Remedies

- Increase DR.
- Increase overfeed in the setting zone.
- Maintain a uniform heater temperature of the texturising heater and improve cooling of the yarn before friction disc unit.

4.2.4 Crimp Rigidity Too Low

Description

- The crimp in the yarn is not rigid and is changing after each application.

Causes

- DR too low.
- Insufficient fixation of twist in the texturising zone.

Effects

- The fitting of the cloth on the body shall not remain the same and is uncomfortable for the wearer.

Remedies

- Increase DR.
- Increase the temperature of the texturising heater.

4.2.5 Surging

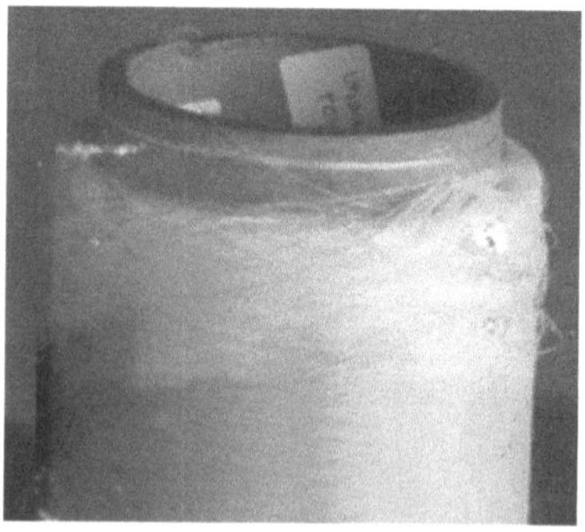

Fig. 4.2.5: Surging

Description

- Instability in the thread line. It gives a lean untexturised appearance.
- Unstable running of the filaments or uneven yarn slippage at the friction disc unit.

Causes

- Uneven yarn slippage at the friction disc unit.
- Large fluctuations in denier of POY.
- Large fluctuations in extension of break of POY.
- Decreased D/Y ratio or tension T1 too low.
- Damaged friction disc unit.
- Speed variation at friction disc.
- Yarn temperature in the texturising zone too high.
- Wrong throughput speed for a given product.
- Irregular application of spin-finish.
- Irregular yarn orientation along the length of the yarn.

Effects

- Uneven yarn.
- Uneven build of the take-up package.
- Slippage of the yarns making it unsuitable for working in the next process.

Remedies

- Reduce the quantity of spin-finish.
- Use spin-finish with increased disc friction.
- Use disc with high yarn grip.
- Reduce the speed of the texturising machine.
- Check feed roller units.
- Change POY package.
- Reduce the texturising temperature and improve cooling of the yarn before friction disc.
- Increase the frequency of cleaning.
- Increase D/Y ratio.
- Check the speed with stroboscope.
- Increase D/Y ratio.

4.2.6 Yarn Comes Out of Input/Intermediate/Output Nip Roller

Description

- The yarn while running on the machine comes out of the gripping rollers, may be input rollers, intermediate guiding rollers or the output nip roller and low nip roller pressure.

Causes

- Traverse is not set properly.
- Distorted yarn guides.
- Vibrations in the machine or in the yarn.
- Low nip roller pressure.

Effect

- Missing effects of texturising in between.
- Uneven shrinkage.
- Uneven crimps.
- Unwanted effects of dyeing on the fabric.

Remedies

- Set traverse with the help of turn buckle provided with reduction gear box.
- Replace with new guides.
- Set the nip roller pressure.

4.2.7 Yarn Not Passing through the Heater Tube

Description

- The yarn does not pass through the heater tube and falls out resulting in breakage.

Causes

- Blocking of the secondary heater tube.

Effects

- Breakage in the yarn and loss of production.
- Sometimes, a broken yarn can interfere with the adjacent good yarns running in the machine or lap on the moving parts of the machines leading to breakdowns.

Remedies

- Remove the foreign particles by cleaning the tube with 10% NaOH solution.
- Check and clean the waste yarn from the suction tubes.

4.2.8 Cutter Not Functioning

Description

- Cutter does not cut the yarn when required.

Causes

- Disturbed detector connection.
- Detector may not be working.
- Cutter faulty.

Effects

- If the running yarn breaks or laps on the nip roller, the yarn is not cut and allowed to pass on.
- When the yarn is not cut, it moves on along with the running materials generating a defect.

Remedies

- Service the cutters periodically.
- Replace the blades if needed.
- Check and correct the detector connection.

4.2.9 Yarn Lapping

Description

- The yarn sticks to the input or delivery rollers and lap on them.

Causes

- Detection fails, that is, broken end detection unit does not sense the broken end.
- Suction may be proper.
- High static charges can also lead to lapping.

Effects

- Lapping is very difficult to remove. Knife needs to be used to remove it. It may damage the machine parts.
- If lapping is not noticed in time, it might lead to fire.

Remedies

- Remove the waste from the yarn collecting tank.
- Keep the surface of the rollers free of sticky coatings.
- Maintain the required temperature and humidity to avoid generation of excess static charges.

4.3 Yarn Damages

4.3.1 Broken Filaments

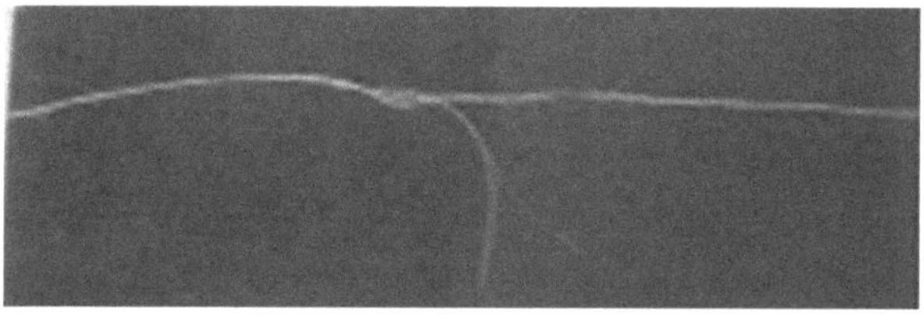

Fig. 4.3.1: Broken filaments

Description

- The filament end protrudes from the yarn bundle as well as from the package.

Causes

- ***Damaged friction disc***: Discs in the friction-twisting unit are responsible for rotational movement in the yarn and, in turn, imparting twist. Damaged disc surface abrades the multifilament yarn vigorously causing surface etching of the polymer leading to generation of broken filament.
- ***Waste on feed roller/bowl:*** The wastes on the feed roller bowl catch the loose filaments and break them.
- ***Higher D/R ratio:*** In multifilament yarn, there are variations in the yarn properties and also between the filaments. When a high DR is applied, some of the filaments reach their breaking extension and break leading to broken filament.
- ***Dirty primary heater:*** Dirt inside the heater acts as an obstacle for the free movement of the yarn, and hence the filaments break.
- ***Thread guide damage/worn:*** Damaged or worn out portions in yarn path catches the filaments and breaks them. As the thread passes through the thread guide, the condition of it should be perfect all the time. Even a slight scratch can result in frequent breakages.

- ***Apron belt damage:*** Damage in apron belt results in erratic movement and uneven twist and tensioning of the yarn resulting in breakages.
- ***POY with low spin-finish***: Uneven spin-finish allows the static charges to generate when the yarn is moving or rubbing with other yarns or metal surfaces. This results in breakage of the yarn.
- Lower D/Y ratio.
- ***Mechanical damage to supply yarn***: The supply yarn which has mechanical damage shall break when the yarns are stretched, as the damage enlarges and becomes big when stretched.
- ***Misthreading:*** Yarn when not threaded properly shall generate excessive tension and breaks the yarn.
- ***Heater temperature too high:*** The high temperature fuses the thermoplastic filaments and softens the filaments resulting in breaks.
- ***Machine speed too high:*** With very high speed of the machine, the tension increases and might cross the limits of the breaking strength. The speed should be set considering the tensile properties of the yarn being worked.
- ***Yarn runs on sides of primary heater:*** When the yarn runs on the sides of the primary heater, some filaments might touch the heating element and get fused or broken, resulting in yarn breakage after stretching.
- ***Yarn runs on sides of cooling grooves:*** When the yarn runs on the sides of the cooling groves, it can get caught in the groves and break.
- ***Cut in nip rollers:*** During texturising, any cut in nip roller retains or holds some of the filaments in multifilament yarn, whereas the other filaments in the yarn maintain the desired linear and migratory movement causing the retained filaments to reach their breaking extension leading to generation of broken filaments or yarn breaks.
- ***Denier is very low:*** When the denier is too low, the breaking strength of the yarn shall be low. Hence, the speed, which is set considering the normal denier, shall develop tension higher than the breaking strength of low denier yarn present in between, resulting in the breakage of the yarn.
- ***Mechanical damage in the thread path:*** Any damage in the thread path shall catch the filaments and prevent the yarn from moving forward. As the take-up rollers are pulling the yarn, breakage shall occur.

Effects

- Stoppages during warping by the fault detector in the warping machine.
- Cause breakages during winding and twisting.
- Increased filament tension on the loom.
- More breakage on the loom.

Remedies

- Check primary heater, thread guide, traverse guide and apron belt.
- Set top–bottom header guides and set the yarn path in such a way that the yarn runs at the centre of the track.
- Set bottom heater guides and cooling plate in such a way that the yarn runs at the centre and approximately 5 mm away from the edges at both the ends of the cooling plate.
- Optimise D/R ratio to be used.
- Operational DR to be a step lower.
- Use nip roller with defect-free cot.
- Friction twisting discs should be in good condition.

4.3.2 Filmentation

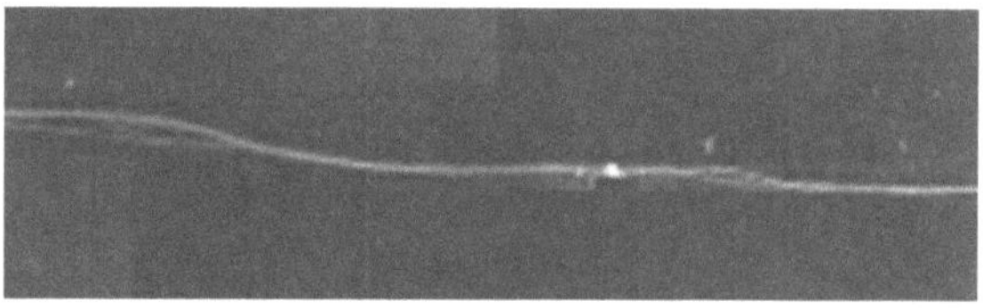

Fig. 4.3.2: Filamentation

Description

- Uneven running off from spinning package, crossed ends.

Causes

- Excessive friction at guides and thread line.
- Insufficient filament cohesion.
- Defective filament guides and incorrect threading.
- Defective friction discs.
- Excessive heater deposits.
- Heater temperature too high.
- Yarn tensions T1, T2 and T3 very high.

Effects

- Frequent yarn breaks.

Remedies

- Reduce D/R at texturising.
- Decrease D/Y.
- Check the speed (RPM) with stroboscope and correct it if deviations are found.
- Check guides and thread liver.
- Use POY with higher pre-orientation.
- Change the number of discs or disc type.
- Reduce texturising machine speed.

4.3.3 Weak Yarn

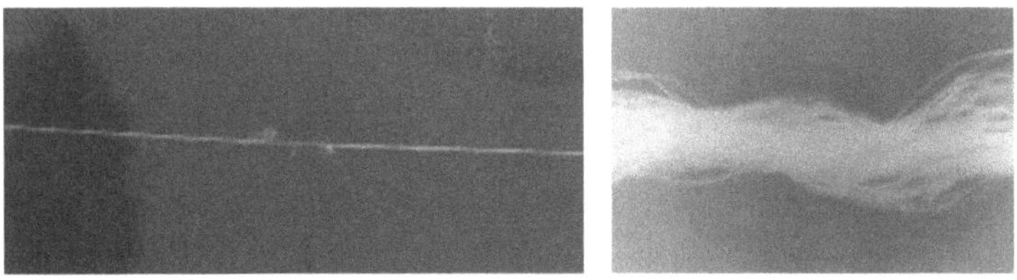

Fig. 4.3.3: Weak yarn

Description

- The yarn having a very low strength and elongation is called a weak yarn.

Causes

- High D/R ratio.
- Primary heater temperature high.
- Poor quality thread guide.
- Variation in elongation.
- T2/T1 not optimum.
- Bad POY package.
- Low POY shrinkage.
- Worn-out friction disc.

Effects

- Increases breakage when working on the machine.
- Reduces production.

Remedies

- Adjust D/R setting.
- Preventive checking of thread guides.
- Optimum primary heater temperature.
- Reduce D/R and D/Y.
- Check tenacity, extension at break and U% of POY.
- Check application of finish.
- Check POY shrinkage (should be around 52%–62%).
- Friction disc should be in good condition. Use of re-coated disc to be avoided. Change of disc should be performed.

4.3.4 Yarn Breaks

Description

- The yarn break rate is the number of yarn breaks that take place per unit weight of the texturised yarn produced. When the breakages cross a specified accepted limit, it is termed as high breakages.

Causes

- The breaks can be caused by low-quality POY.
- Excessive high DR.
- Excessively high primary heater temperature.
- High processing speeds causing instability and surging in the thread path.
- Poor maintenance.
- Worn or damaged ceramic discs.
- Poor thread path alignment.
- Damaged nip rolls.
- Damaged aprons.

Effects

- Yarn breaks can create short yarn packages, loss of production and increased labour and packaging costs.

Remedies

- Low-quality POY should be segregated.
- Reduce machine speed.
- Monitor the maintenance activities.
- Have periodic machine audit and identify the spindles giving repeated breaks.

4.4 Winding Defects due to Process Variations

4.4.1 Poor Rewound Package

Fig. 4.4.1: Poor rewound package

Description

- Winding of the yarn on the paper tube not satisfactory.

Causes

- Stroke length disturbed.
- Low hardness of the wound package due to low tension.

Effects

- Unwinding problem shall be there in the next process. The yarn shall break frequently.
- Tension variations shall be created while knitting resulting in barre effect.

Remedies

- Good stroke length setting.

4.5 Winding Defects due to Setting

4.5.1 Jali (Open Mesh-Type Appearance)

Fig. 4.5.1: Jali

Description

- Open net-like or mesh-type appearance of overthrown ends on the side of the package is called webbing, popularly known as 'jali' in India.

Causes

- ***Low take-up overfeed:*** Yarn tension in the take-up zone rises with lower take overfeed. The high take-up tension disturbs the laying of yarn leading to yarn slipping from the position and appears as overthrown ends in the package and appears as jali.
- ***Cradle vibration:*** Vibration in the cradle erratically shifts the winding point leading to misplacing of the yarns in the form of overthrown ends and appears as jali.
- ***Crossing angle disturbed***: The yarn slips out at the reversing position creating a jali effect.
- ***Traverse guide damaged:*** A damaged traverse guide may make erratically excessive or lower movement resulting in cross ends and a jali appearance.
- ***Damaged paper tube***: Paper tube jump due to damage disturbing the layers and creating a jali.
- ***Paper tube not held tight***: Loose paper tube jumps and create mesh-like formation.

Effects

- Unwinding problems.
- Frequent breakages while unwinding in the next process.
- More yarn wastes as the jali portion needs to be cleaned and fed.

Remedies

- Increase take-up overfeed.
- Adjust cradle spring provided at the base to stop vibration.
- Check crossing angle, traverse guide and paper tube.
- Reject damaged tubes.

4.5.2 Bulging

Fig. 4.5.2: Bulging

Description

- Wound yarn layers projecting out of normal position.

Causes

- ***Damaged paper tube:*** Paper tube jumps due to low take-up cradle pressure while working and the pressure on the package shall be less resulting in loose wound package at that point.
- ***Disturbed stroke length***: When stroke length is disturbed, at certain places the yarn wound shall be less compared with other places resulting in uneven pressure on the package being wound. Bulging takes place wherever the pressure is less. In the portion where the yarn winding is more, the package becomes tight and pushes the adjacent portion to form a bulge.

Effects

- Unwinding problem.
- Chances of breakage.

Remedies

- Adjust the cradle pressure.
- Check paper tube quality and reject tubes of poor quality.
- Check the stroke length and correct if necessary.
- Do not use damaged paper tube.

4.5.3 Slippage

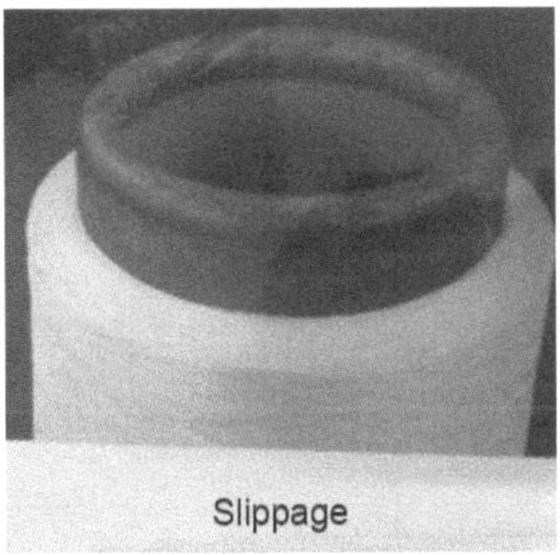

Fig. 4.5.3: Slippage

Description

- Texturising effect not obtained due to slippage of the filament during texturising

Causes

- Apron belt nip roller disturbed or damaged.
- Damaged friction discs.
- Higher D/Y ratio.

Effects

- Yarn without bulk.

Remedies

- Remove worn out aprons, nip roller with groves and damaged discs.
- Correct the D/Y ratio.

4.5.4 Tight Winding

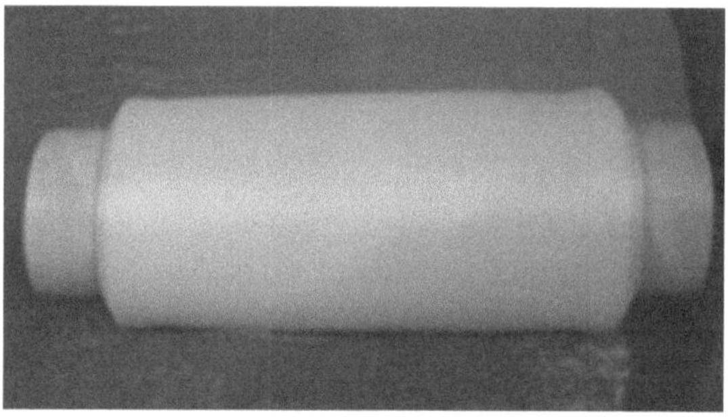

Fig. 4.5.4: Tight winding

Description

- Very compact yarn package or yarn wound with excess tension.

Causes

- Winding speed is higher than the requirement considering the yarn elongation properties.
- The tension of winding is more than the required as per the tensile and elongation properties of the yarn.
- Taper angle disturbed.
- High take-up cradle pressure.

Effects

- The packages become hard leading to tension variation while unwinding and cause fabric defects.
- Unwinding problem in twisting process.

Remedies

- Checking the tensile properties of the yarn and adjusting speed and tension suiting the yarn quality.
- Checking and setting of taper angle.

4.5.5 Bad Build-Up

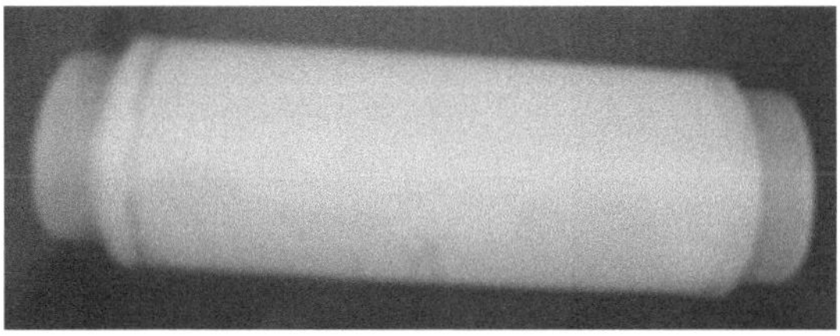

Fig. 4.5.5: Bad build-up

Description

- Any deviation of the package from the cylindrical shape is called bad build-up.
- Overthrown ends, ridges, shouldering and no-tail packages are other common package build faults.

Causes

- ***Misaligned cradle:*** The yarn package experiences uneven pressure exerted by the cradle when it is misaligned with the winding drum, leading to deshaped package.
- ***Yarn not running through proper bar guide:*** The winding tension rises excessively coupled with the rise in tension at the reversal point of the traverse guide lead to deshaped package.
- ***Small wind angles:*** One of the more common package build fault is bulging, caused by using small wind angles.
- ***Webbing or scrambling:*** Webbing or scrambling of the yarns on the package many times caused by a too low wind angle for intermingled products.
- ***Density:*** High-density packages produced with high take-up tensions will also result in webbing.

Effects

- Poor yarn quality.
- Unwinding problem.

- Production and efficiency reduce in the next process and customer complains.

Remedies

- Adjust the cradle alignment.
- Yarn path alignment for a uniform yarn tension.

4.5.6 Stroke Length Disturbances

Fig. 4.5.6: Stroke length disturbances

Description

- The variation in the length of yarn wound on the package.

Causes

- Take-up tube loose on end caps.
- Misaligned cradle.
- Stroke length setting disturbed.
- Traverse guide loose/worn.
- Traverse guide has some damage or wear.
- Point of origin guide not in centre.
- Package bouncing.
- Cradle catching on feedback arm.
- Take-up tension too high.
- Winding angle too low.
- Taper angle too low.
- Radial disturbance too high.

Effects

- Higher sub-standard quality of the yarn and loss.
- Problem during unwinding.

Remedies

- Stroke length adjustment.

4.5.7 Improper Shore Hardness of the Package

Description

- The wound yarn packages have different Shore hardness between the packages and not as per the agreed norms.

Causes

- The take-up cradle pressure is not proper.
- The winding tension is not uniform.
- The winding tension set is either too low or too high for the yarn being worked.

Effect

- The Shore hardness of the yarn package is very important for getting uniform tension while unwinding.
- If the Shore hardness is more, the yarn shall be tight whereas if it is low, sloughing off may take place while unwinding.

Remedies

- Maintain a proper winding tension and the take-up cradle pressure.

4.6 Defects due to Damaged Parts in Machine

4.6.1 Damaged Disc

Fig. 4.6.1: Damaged disc

Description

- Disc surface broken.

Causes

- Bad operating during cleaning.
- Cutting problem.
- Over life.

Effects

- Broken filament.
- Threading problem.
- Untwisted yarn.
- Lower yarn quality.

Remedies

- The cutter should be checked.
- Good operating.
- Good threading.
- Ceramic discs are to be stored properly in dry area as they are very brittle and sensitive to various mechanical actions. Care should be taken at the time of
  - Knocking and dropping of the discs.
  - Exposure of dropping discs to heat/cold.
  - Cleaning of the discs in an ultrasonic cleaning bath is recommended.

4.6.2 Damaged Thread Guides

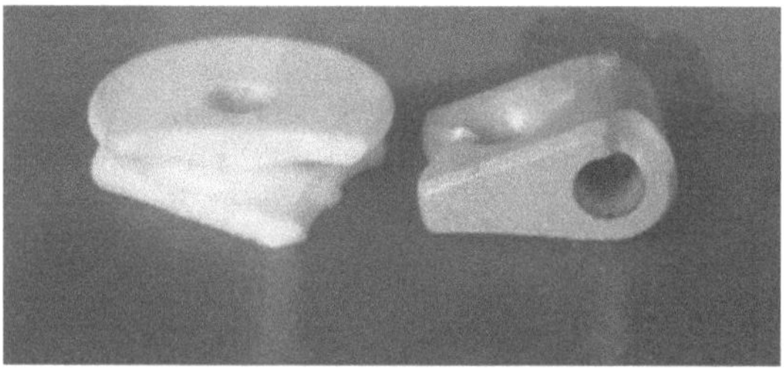

Fig. 4.6.2: Damaged thread guides

Description

- The thread guide not in proper shape or broken.

Causes

- ***Rough handling:*** The thread guides are made of fine ceramic; any hard materials hitting the thread guides break them.

- ***Threading mistake by operator:*** When the threading is wrong, the tension increases on the yarn which, in turn, acts on the thread guide. As the synthetic yarns are very strong, sometimes it results in pulling the thread guide out or breaking the thread guide.
- ***Poor quality of thread guides installed***: If the quality of the guides is not good, even a slight increase in load can break them.
- ***Higher guide life:*** The ceramic guides have certain working life. If it is used for more than that specified life, the guide becomes brittle and gets damaged.

Effects

- The filaments break if the thread guides are cracked or worn out.
- Winding disturbance.

Remedies

- Analyzing the reasons for the thread guides getting damaged and taking corrective actions.
- Checking the condition of the thread guides periodically during maintenance and replacing the broken or worn parts.
- Experienced operators to be engaged who can detect bad parts fast and alert the fitter.
- Good supervision.

4.6.3 Damaged Disc Driving (Positive-Torque) Belt

Description

- The disc driving Positorq belt worn out or torn out partially.

Causes

- Belts distorted or stretched.
- Oil contaminating with the belts.
- Operator trying to slow the spindle by pressing it by hand.

Effects

- Results in improper disc movement.
- Worn teeth cause the belts to jump pulley teeth; as a consequence, there is loss in spindle speed and invariably inferior texturised yarn.

Remedies

- Disc driving belts have glass reinforcing cords, which can damage if distorted or stretched. The belts should be stored on pegs, tubes or loosely on edge in shallow trays avoiding heavy object resting on them. They should be kept dry and free from oils, solvents or abrasive material.

- Ensure that the yarn wraps do not build up and consequently seize a spindle
- Clean snow accumulation on pulley teeth at regular intervals.

4.6.4 Destruction of Gear Belts

Description

- The gear driving belt has become loose or worn out.

Causes

- Jamming of friction disc due to lap formation or deposits unsuitable for spin-finish which leads to the destruction of the gear belts.

Effects

- The drive to gear gets disturbed leading to more breakages of the yarn while the machine is working.
- The broken belt may entangle and damage other parts of the machine.

Remedies

- Follow maintenance schedule strictly.
- Have a system of periodic checking of machine parts and belts by conducting machine audits.
- Get the belts from reliable source only.
- Store the belts in open form to avoid deformation due to heat.

4.6.5 Bearing Damage at Friction Disc Unit

Description

- The friction discs are not moving freely due to bearing damage.

Causes

- Longer period of stoppages of friction unit.
- Fluff and dust accumulation.
- Lubrication drying up.
- Periodic cleaning of the spindles is not done as per schedule.
- Poor quality lubricant used.
- Improper alignment of friction disc spindles.

Effects

- When the disc unit revolves slowly or stops in between, the twist inserted shall get disturbed leading to uneven texturising.

Remedies

- Stop friction units for short periods only.
- Dismantle friction units with bearing – lift from support.
- Strictly follow the schedule of maintenance and cleaning of the spindles.

4.6.6 Excessive Vibration/Noise from the Feed Shaft

Description

- Excessive vibrations and noise observed from the feed shaft.

Causes

- Shaft alignment is not proper.
- Bearing runs dry.
- Bearing failure.
- The shaft is worn out at bearing place.
- The bearing seat is worn out in the bearing holder.

Effects

- The vibrating shaft gives irregular feed resulting in periodic variations in the yarn.
- The yarn U% shall increase which can give patterns on the fabric.

Remedies

- Check it and reset if required.
- Periodic maintenance of the bearing mounting the shaft.
- Check the shaft for balancing.
- Replace the bearings if worn out.
- Replace the roller stand/bearing holder.

4.6.7 Hairy Yarn

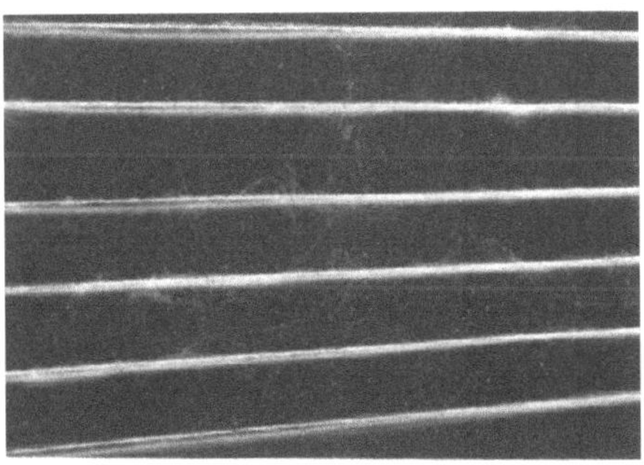

Fig. 4.6.7: Hairy yarn

Description

- In multifilament yarn, breakage of substantial amount of the filament causes its ends to protrude from the surface as well as from the package and has distinct 'furry' appearance known as hairy.

Causes

- ***Damage to filament while in process:*** In a filament there shall not be any hairy portion. Due to any reason, if a filament is broken, the broken portion comes out of the yarn and gives the appearance of a hair.
- ***Damaged groove guide:*** The groove guide is situated in between the twisting assembly and the intermediate roller. The yarn makes two and half rounds on it. Because of the positional effect, the yarn exerts substantial pressure on it. Even a slight roughness on the guide surface causes appreciable number of filament breaks, which appear as hairs.
- ***Misaligned cooling plate:*** The positional adjustment of the cooling plate is such that the yarn grazes over it during operation. The yarn experiences substantial amount of friction when it is misaligned causing appreciable number of filaments to reach their breaking extension leading to hairy yarn.
- ***Low POY shrinkage:*** Low POY shrinkage signifies high amorphous and low crystalline orientation. Consequently, POY of this type possesses extensibility than the normal POY. When it is subjected to draw applicable to normal POY, a substantial number of filaments reach their breaking extension quickly resulting in hairy yarn. However, the texturising unit cannot check this.
- ***Jammed feed nip roller:*** Damaged bearing of the nip roller hinders free rotational movement of the yarn. When the feed nip roller is unable to rotate freely, the effective DR increases causing an appreciable number of filaments reaching their breaking extension leading to hairy yarn.

Effects

- Increase in poor quality yarn.
- Frequent breakage of the yarns while working on the machine.
- Reduced production at the next process.
- Increase in waste level.
- Barre and hazy appearance of the fabric.

Remedies

- Condition of the groove guide should be reviewed periodically.
- Cooling plate as well as the whole yarn path to be checked on schedule basis.
- POY having shrinkage between 55% and 65% to be used.
- Each part in the nip roller assembly to be checked thoroughly during maintenance schedule.

4.7 Defects due to Machine Setting

4.7.1 Speed Variation at Friction Disc Unit

Description

- The speed of the friction disc unit changing inadvertently.

Causes

- Damaged bearing driving the spindle.
- Lapping at the bearing or yarn entering the bearing.
- Faulty drive belt or belt damaged.
- Dirty or oily belt.
- Excessive wear of spindle wharve.
- Defective bearing.
- Wharve does not run properly.

Effects

- Bulk variation.

Remedies

- Check RPM of the friction disc by stroboscope and take appropriate action.
- Strictly follow the maintenance schedule and replace worn-out parts in time.

4.7.2 Top/Bottom Heater Guides Not Moving with Detent Mechanism

Description

- The heater guide, either top or bottom, is not moving with detent mechanism.

Causes

- Half coupling is loose or broken.

Effects

- Leads to uneven heat-setting and heat damages in the yarn.

Remedies

- Tighten if loosen.
- Replace if broken.

4.7.3 Positorq Unit Stops

Description

- Positorq unit driving the discs stops working.

Causes

- Yarn entering the bearing.
- Positorq driving belt damaged.

Effects

- Twist shall not take place when the drive to the discs is stopped, and hence there shall not be texturised effect.

Remedies

- Remove the unit and clean out the yarn.
- Replace the belt with a new one.

4.7.4 Slippage between Positorq Wharve and Drive Belt

Description

- Slippage found between Positorq wharve and drive belt.

Causes

- Dirty or oily belt.
- Excessive wear on wharve.

Effects

- Slippage results in the variation in twists, leading to the variation in crimp and loop height.

Remedies

- Tighten the belt by adjusting the tension guide.
- Replace the Positorq belt (disc driving belt) if necessary.

4.7.5 Noisy Positorq

Description

- The disc driving belt unit is making noise.

Causes

- Defective bearings in the belt driving unit.
- Dried out bearing units.

Effects

- Noise is an indication of the bearing being worn out. It shall lead to an uneven drive to the disc unit resulting in uneven twist, uneven crimp and uneven loop heights.

Remedies

- Replace the bearings if found worn out.

4.8 Defects due to Human Work Practices

4.8.1 Dirty or Oily Package

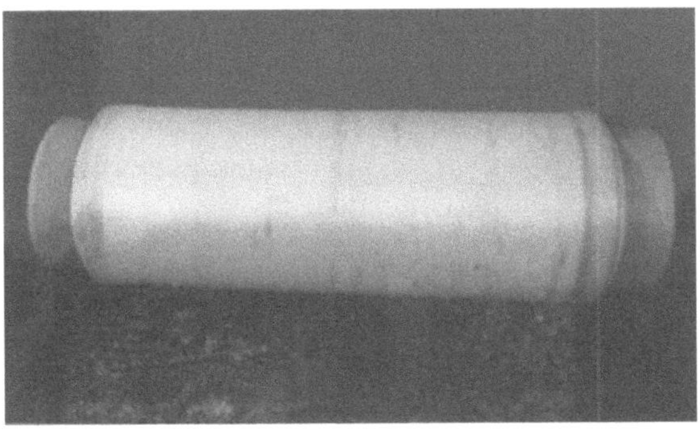

Fig. 4.8.1: Dirty or oily package

Description

- The texturised yarn having dirt and oil on it.

Causes

- Poor housekeeping and oil spilling on the floors and from the machine parts.
- Poor handling by the operator.
- Lack of skill and knowledge of the operators.

Effects

- Downgrading of the yarn resulting in low price realisation.
- Staining in the fabric.
- Breakage at the loom or knitting machine.

Remedies

- Practice good material handling and housekeeping systems.

4.8.2 Poor Storage

Description

- The texturised yarn packages are not stored in order.

Causes

- The conditions of storing not proper.
- Package kept exposed to open atmosphere, moisture and dust.
- The materials produced are not packed in time and stock has increased in the racks.
- Insufficient number of the racks for the rate of production and packing capacity.
- The workers are not educated on the harm of improper storing.

Effects

- A change in colour or shade.
- Decreased lustre.
- Oil will dry and forms coloration.
- The coils get disturbed and results in more wastes.

Remedies

- The conditions of storing should be proper.
- The package should be properly kept in the rack.
- The supervisor should visit the packing area periodically and monitor storing.
- Follow first-in-first-out system while sending the materials for packing.
- Educate the workers again and again regarding good storing practices.

4.8.3 Wrong Threading

Description

- Improper threading of the filament.

Causes

- Bad operating conditions.
- Untrained workmen working on the machines.
- The supervisor not inspecting the machines properly.

Effects

- Untexturised yarn.
- UDY.
- The untexturised yarn without oil.

Remedies

- Good supervision.
- Training of workmen.

4.8.4 Damaged Package

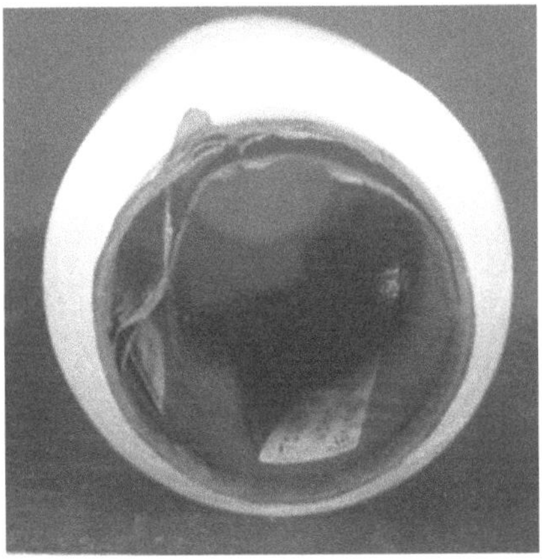

Fig. 4.8.4: Damaged package

Description

- Damaged packages.

Causes

- The quality of the paper tube used is not proper.
- Poor handling of the packages.
- The packages coming in contact with water (roof leakages or water seeping in the floor).
- Poor supervision and lack of education to workers.

Effects

- The yarns cannot be used if the packages are damaged.
- The yarn wastes shall increase.

Remedies

- Educate the workers and supervisors on the losses due to package damages.
- Provide proper and adequate tools for handling packages.
- Procure and use only good quality paper tubes.

Fig. 4.8.4.1: Pegged trolley

- Use of peg trolley for transportation can reduce damages to the packages.

4.8.5 Poor Package Unwinding

Description

- Unwinding of the texturised yarn from the package not proper.

Causes

- Stroke length incorrect.
- Cradle setting incorrect.
- Filmentation in the texturised yarn.
- The yarn coming out of the traverse guide.
- Take-up tension too low.
- Winding angle too high.
- Taper angle too low.

- Radial/axial disturbance link out of phase.
- Incorrect oil.
- The package webbed.

Effects

- Higher generation of hard wastes.
- Low production in the next operation.
- The yarn gets downgraded and price realisation shall be low.

Remedies

- Check the setting of the stroke length of the traverse guide.
- Check the cradle setting.
- Keep proper take-up tension.
- Reduce winding angle.
- Check taper angle.

4.8.6 Oil Spot

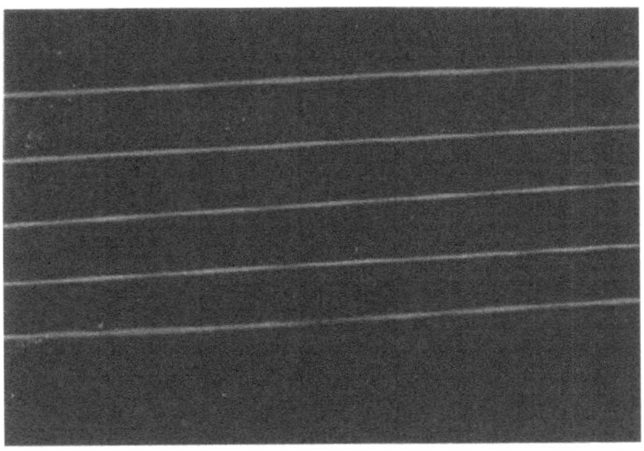

Fig. 4.8.6: Oily spot

Description

- Oil stain on the yarn and package.

Causes

- ***Leak in the oil-sealed bearing of cradle end cap:*** The end cap in the cradle is fitted with oil-sealed bearing. The leakage in the bearing seal lead to oil splitting on the bobbin remitting in oil spots.
- ***Dripping of lubricant oil from cam-box***: Excessive oil in the cam-box drips down on the bobbin underneath it causing oil spot.

Effects

- Ugly surface look of the yarn as well as fabrics produced out of it.
- The oily portion does not absorb the dyes and hence gives a shabby look of the fabric.

Remedies

- Use authentic end cap bearing.
- Thorough checking to be done during scheduled maintenance.
- Care to be taken during cam-box lubrication to avoid excess oil filling.
- Ensure that the hands of the operator are clean before handling the materials.
- Such bobbins should be downgraded during the inspection and packing.

4.8.7. Without Oil

Fig. 4.8.7: Without oil

Description

- The yarn winding without oil.

Causes

- The yarn does not touch the oil roller.
- The oiler roller thread guide disturbed.
- No oil in the oiler tray.

Effects

- Low denier.
- Breakage at the loom due to high static charges.

Remedies

- Setting the oil roller properly.
- Good supervision by the supervisors.

5

Towards Defect-Free Product and Services to Customer

In the previous chapter, we discussed on the defects, their causes, effects and remedies in the texturised yarns; however, it helps only for taking corrective action after a defect has been produced. It is adding cost to the product. A poor quality produced hampers production activities, reduces the production output, increases wastages, delays delivery to the customer and reduces the confidence of the customer in the organisation. A good yarn manufacturer shall work for producing the required quality yarn in the first attempt itself and shall work for maintaining that quality consistently. It is called as Quality Management System (QMS).

To implement good QMS, a whole hearted support of all the workmen, staff and management is needed. Achieving whole hearted support is possible when people appreciate the requirements, accept it, willfully work to achieve it and enjoy their work. It is termed as Work Quality Management (WQM).

To have this right thing happening right at first time and at all the times, there should be QMS and WQM at all levels and processes of the organisation, which is termed as Total Quality Management (TQM).

5.1 Work Management

To get the required quality and productivity, one need to plan the activities, acquire the required raw materials, assign the suitable machines, decide on the standard operating procedures and controls to be exercised, train the operators to do the work and supervise the activities meticulously.

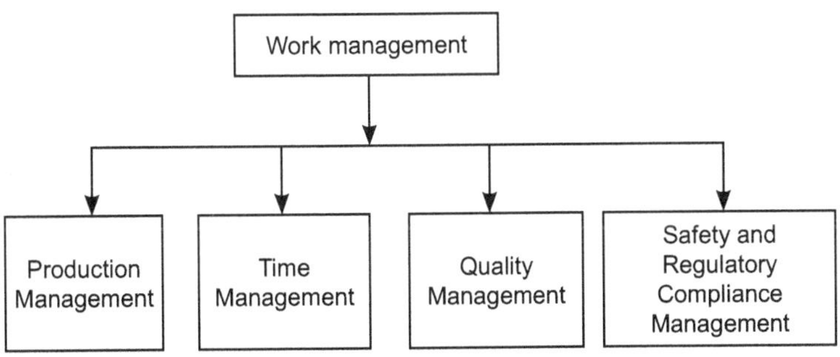

Fig. 5.1: Components of Work Management

The Work Management has four major components, namely, Production Management, Time Management, Quality Management and Safety and Regulatory Compliance Management.

5.1.1 Production Management

This deals with getting the required production by deciding the process parameters, speeds, production efficiency of each machine and operation, planning the required machines, balancing the available machines considering their efficiencies and capabilities, maintenance schedules, maintaining the process, maintaining the machines, assigning the responsibilities and supervising the production and process flow.

Timely indenting of the raw materials, spares and packing materials also becomes a part of Production Management.

Monitoring the production on hourly basis is very important to achieve the targeted production. The production management also involves the monitoring of idle production positions and stoppages. Starting the machines without delay after each doffing operation, lot changing or maintenance is very important to achieve the productions.

The production management also involves allocating works to suitably trained people, giving clear instructions and supervising their works to ensure that the production is achieved as needed.

Improper production management results in shortage of feed materials, unwanted process stocks waiting for further process and more wastes due to the materials getting deteriorated or accumulating fluff and stains. When the materials run out without supply of sufficient materials from back, the package sizes get varied and also the tension on wound coils.

5.1.2 Time Management

To adhere to the deliveries in time, one has to plan the activities and their sequence after understanding the time required for each operation. Studying and understanding the time required for each job, understanding the priorities of the activities, preparation of process flow chart and assigning time for each process are essential. Grouping the activities as ‘urgent’ and ‘not urgent’ and as ‘important’ and ‘not important’ helps in prioritising the activities. The activities are grouped further as follows which give clear idea as to what is to be done first.

1. Urgent and important,
2. Important but not urgent,

3. Urgent although not important and
4. Neither important nor urgent.

Understanding the supply chain, lead time required for procuring each item, placing orders in time and getting the materials in time, understanding the availability of workmen and arranging for them in time, planning and following up with the logistic arrangement are all components of time management in a yarn texturising factory. Meticulous follow-up starting the activities in time and analyzing the delays and taking suitable remedial actions help in achieving the deliveries in time. Developing the culture of commitment to the agreed time, starting the activities in time and monitoring the activities where time loss is found affect the complete organisation.

In time management, it is necessary to identify the ineffective times which do not result in production, quality or a profit to the organisation and such activities need to be replaced with effective activities. Unwanted activities and duplication of activities are to be eliminated. This reduces the stress on people and allows them to concentrate on production and quality in a better way.

Improper time management results in improper supply of the material in time for different activities, stoppages of the machines waiting for the materials, may be raw materials of empty take-up packages or packing materials, stickers and so on. This results in poor efficiency and also poor quality.

5.1.3 Quality Management

Quality management addresses getting the required quality at the first time by devising suitable method of working to get the required quality, documenting the procedures and providing for the people working, deciding on the quality of raw materials and process parameters, training and educating the workmen to do the working, preparing and providing work instructions, check sheets for checking and ensuring the quality of the materials received, the quality of the material in process and the quality of the materials produced. Monitoring the activities at all levels is essential to identify the deviations in process and in quality and take suitable corrective actions in time and preventing the poor quality from moving forward is very important. Periodic auditing of the processes and the systems to ensure that everyone has understood and is following the procedure exactly as per requirement is very part of a QMS. Proper implementation of a suitable QMS helps in achieving the required quality and consistency in quality.

Quality management has four components, namely, quality planning, quality control, quality assurance and quality improvement.

- Quality planning addresses planning of all activities considering the specific requirements of the customers. It involves planning and allocation of the machines, planning and acquiring of raw materials and spares, planning of workforce and providing them, scheduling the activities of production and maintenance and planning the controls to be exercised at various levels of production.
- Quality control ensures that poor quality is not delivered to the customer and poor materials are not mixed with good materials. It involves testing of the incoming raw materials, auxiliaries and spares, the materials in process and the materials produced. The materials which are found to be of nonconforming quality to the norms are rejected and prevented from getting along with good materials. It also involves verification of testing and monitoring devices and getting them calibrated in time. Suitable statistical techniques are adapted for deciding on the sample size, the number of testing to be done, periodicity or frequency of testing and acceptance criteria for the quality of materials and for passing the materials.
- Quality assurance addresses the method of ensuring that all activities are done as per agreed procedures to get consistency in quality all the time. It involves development of standard operating procedures, work instructions, check sheets, process parameters, reviewing the activities, auditing the systems and taking corrective actions wherever a deviation is noticed. It also involves providing the required resources for doing the activities, educating and training the people doing the work to ensure quality and proper allocation of people with the required competency for the jobs being done.
- Quality improvement involves identifying the potential poor quality or problems and taking proactive corrective and preventive actions, and making the system robust. Collecting feedbacks on the quality and service by both external and internal customers, analyzing the reasons for deviations or defects, the exact requirement of the customer depending on the end uses and devising methods and implementing the new decisions become a part of improvement activity. Development of teams and ensuring team cohesion is a very important aspect of quality improvement and sustenance.

Procedure: Procedures indicate the method of doing a work to achieve a specified objective, the people involved in it, the responsibility of each in making the activity fruitful, the factors to be controlled, operating checks to be done, recording and reporting of the activity. Developing procedures is the first step of getting success. If procedures are appropriate to the activity and the objectives, we are certain to get the results provided we follow it religiously. The procedures need to be established before documenting and ensured of following religiously. The interactions between activities in a process should be clearly explained in the procedure.

Work Instruction: Numbers of small activities are involved within a procedure. The person doing that activity should be clear of that work. This should be documented with illustrations or pictures and given to the worker so that he/she can refer and do the work. This is termed as the work instruction.

To do any work successfully, we should be clear about the method of doing work, the sequence of activities, the breakup of activities and distributing to different people in the team, the controls and checks at appropriate places, coordination of activities of individuals and combining them and finally reviewing and reporting. Whatever may be the experience and knowledge one has, it is not possible all the time to remember everything and follow all the steps. To avoid the misses, and also to help juniors to develop, it is necessary to document the procedures and insist people to read it on regular basis religiously.

Any document of a procedure should first explain the purpose and objective of doing that activity. It should clearly specify the result expected out of that activity. After specifying the purpose and objectives, we need to define the scope of operations, the area in which these operations shall take place and the people responsible for doing different activities within these operations.

Once the purpose, scope and responsibility are spelt out, it is better to explain the importance and the principles of operations, and the technical information in case of operating machinery so that people can be clear about the various functions available in the system. Afterwards the works to be done should be explained step by step so that the persons working do not miss a single step.

In the procedure, write clearly as what is to be done, when it is to be done, where it should be done, how it is to be done, who should do it; why is already in the purpose.

It is very important to have a proper system to identify the defective materials and trace the origin of defects. Unless we identify the source of defects, it is not possible to eradicate them. Hence, there should be a system of labelling each and every product at all levels of manufacturing which should be visible and unique to individual product. The labelling on individual package should help in tracing the exact position of the spinning or texturising machine, the date and time of production. Once the production position producing defect is identified, all the production of that position can be segregated and prevented from getting mixed with good material.

Quality assurance of the texturised yarn product can be broken down into three distinct parts, namely, on-line process control, off-line yarn testing and finally through visual inspection of the package prior to packing and despatch. Process control is carried out in the manufacturing environment itself, whereas yarn testing is done in suitably equipped laboratory at controlled conditions of temperature and relative humidity.

Process control can be defined as the constant auditing of the parameters on the texturising machine. This is an active control system and consists of continual checking of production parameters on the machine to ensure that no parameter drifts outside the specific range. This includes checking of speeds of various parts, visual check of integrity of yarn thread path, pressure and quality of compressed air, routine monitoring of thread-line tension, correct POY or polymer feed-stock, heater temperatures, noise and vibrations, worn-out parts and so on. On modern machines, some parameters are self-monitored by using programmable logics, which include speeds, temperatures, pressure and so on. The system gives an alarm when the specified parameter crosses the set limits. However, the alertness of the people working to the signals given by controls is very important.

The people working should be clear about the control points and check points and regularly-follow them. Each mill or factory has to decide on the control points and check points depending on the quality requirements. The following is a general guideline.

Control Points

1. Selection of product parameters such as the material in the yarn, denier and number of filaments, twist/crimp to be inserted, the size of feed packages and the number of packages needed for the order in hand.
2. Selection of process parameters such as speed, tension, feed and delivery ratio, threading of yarn in friction discs in case of false-twisting, temperature to be maintained in heat-setting unit, air nozzle design and pressure of compressed air in case of air texturising, the delivery speed and so on.
3. Delivery package identification, that is, points to be printed in the label.
4. Deciding the skill requirement and engagement of the skilled employees.

5. Deciding the work norms and allocation of workmen.
6. Designing the maintenance activities and implementing them.
7. Deciding and entering the required information in the log book such as the machines allotted for different deniers, the number of spindles (production positions), the total production achieved in each denier, problems faced, actions taken, instructions for next shift and so on.
8. Cleanliness at all places.

Check Points

(a) Material Related

1. Check the raw material, denier, lustre, the number of filaments, the extent of orientation in POY, spin-finish and cross section.
2. Whether the feed material is as per plan that is, denier, the number of filaments, the material (polyester, nylon, etc.), the number of packages received and the weight of each package.
3. Check the lot number or merge number of the material received.
4. Check the labels on each package and colour codification if any.
5. Check MSDS and the safety instructions.

(b) Texturising Machine Related

1. Whether the creel pegs and the yarn guides are aligned properly.
2. Whether the ceramic guides and tension pins in yarn path are in good condition.
3. Whether the friction discs are in proper condition.
4. Whether the air jet nozzles are of correct size in case in case of air texturising.
5. Whether the temperature indicator and pressure gauges are working properly.
6. Whether the stop motions and signals are functioning properly.
7. Whether the cooling fans, if provided, are properly functioning.
8. Whether the yarn path and the machine are in clean condition.
9. Whether the brakes are functioning properly.
10. Whether the empty tubes are in good condition. Reject crushed, nicked or cut tubes, especially on the nose end.
11. Check for the correct fitting of empty take-up tube on the tube holder.
12. Check the take-up package holder and winding units for vibrations.

(c) Setting Related to Process Parameter Data Sheet

1. Check for machine parameters such as
   - type of heater,
   - length of heater,
   - form of heat transmission,
   - feed system,
   - type of false-twisting mechanism used,
   - type of spindle,
   - type of friction discs
   - package building, mechanism.
   - speed of production,
   - speed of twisting,
   - temperature of heat-setting,
   - time of dwell for the material in heat-setting area,
   - cooling temperature,
   - draw tension,
   - tension in the heater zone,
   - direction of twist,
   - twist level,
   - rate of feeding,
   - DR
   - D/Y ratio
2. Whether the packages are creeled as per programme.
3. Whether the temperature set in heat-setting unit is as per plan.
4. Whether the temperature set at hot draw pin in case of air texturising is as planned.
5. Whether the speed of the machine (feeding, delivery twisting, etc.) is as per plan.
6. Whether the take-up package dimensions set are as per plan.
7. Whether the tension set is as per plan.
8. Whether the yarn tension is maintained uniformly on all ends.
9. Whether overfeed set is suitable to the effect required in air texturising.
10. Whether the wetting arrangement is adequate for the denier of the yarn in air texturising.
11. Whether the choice of jet matches with the overall denier, filament denier and the speed of the process in air jet texturising.

(d) Performance Related

1. Check the breakages to ensure whether the breakages are within norms.
2. Observe the breaks for different reasons and their position and frequency.
3. Whether the machine is giving the required production.
4. Whether the machine is giving the required efficiency.
5. Whether the quality is uniform on all packages.
6. Whether the take-up dimensions and density are uniform and as per requirement.
7. Check the produced yarns for filmentation problem and winding problems. If ridges or a groove greater than 1/8 inch high or deep is found, reject them.
8. Check for tails in the wound package. If there are multiple tails, then correct if possible. Otherwise reject to rewind.
9. Check the produced materials for shade uniformity. No shiny, very dull or shade variation is accepted.
10. Check the materials for slub, loops and kinks.

(e) Documentation Related

1. Whether the labels are entered with all relevant information.
2. Whether the package weights are uniform and as per calculation.
3. Whether the texturised materials were sent to the next process with proper memo and documentation.
4. Whether the remnant supply packages are counted, weighed, recorded and returned to warehouse with proper memo after running out of a lot or completing the orders.

(f) Work Practice Related

1. Whether the workers are following the material handling systems as per requirements.
2. Whether the operators are putting the correct type of knot/splice while mending breaks. (This does not apply for POY and texturising.)
3. Whether the unused materials are removed in time from time to time.
4. Whether the machines and the surroundings are kept clean all the time or not.

5. Whether the operator is keeping his hands clean before touching the material.
6. Whether the worker removes the entangled yarns on package holder before putting a new take-up tube.

(g) Log Book Related

Check whether the following information are entered in the log book:

1. The machines allotted for different orders, the total number of supply packages used, the twist per metre, the operator engaged and the production achieved.
2. Stoppages with reasons for stoppages.
3. The problems faced such as short of materials, materials not received in time and quality-related issues.

(h) Management Information System (MIS) Related

Check for the correct entry of the following data in the system:

1. Order number.
2. Machine number,
3. Number of production positions worked.
4. Nominal count (denier and number of filaments).
5. Weight of supply packages.
6. Weight of delivery packages.
7. Date and time of production.
8. Date and time of issue to the next process.
9. Remnants returned to warehouse.

(i) General

1. Whether the temperature and humidity are maintained as per requirement.
2. The quantity of hard wastes generated.
3. Whether the hard wastes generated are within the norms.
4. Whether the machine cleaning was done properly by the maintenance people.
5. Whether you took charge of the shift in a good condition.
6. How was the condition of your section when you handed over the shift to the next person.

Display the instructions to the operators in the form of Dos and Don'ts with the symbols of ✓ and ✗ suitably.

The aim of the quality assurance system must be to ensure that the physical and dyeable properties of the yarn are those that will enable it to perform satisfactorily in its destined end use. Circular knit, warp knit, woven goods (both as warp and weft) and off course automotive-type products all have their own specific requirements. J. W. S. Hearle et al. have classified the yarn into six major groups considering the similarities in quality requirements. They are as follows:

- Automotive yarns
- Weaving warp yarns
- Weaving weft and circular knit yarns
- Printed fabrics
- Whites only
- Off-quality

It is suggested that each yarn processor collects the exact requirements of the quality from their customers depending on the performance requirement of the end product at their customers' place and categorise the yarn in different categories.

5.1.4 Safety and Regulatory Compliance Management

One can invest huge amount and buy latest machines; but without suitable person to handle it, it becomes a loss (waste). The success of a company depends on the efficiency of the workforce. There is a close relationship between the safety measures and the efficiency of the workers. Efficiency results in increasing the average output per worker which reflects in increased productivity. Hence, ensuring the safety of the people working and also of the assets is very important for any factory. Providing adequate measures for safe working, defining and managing a good housekeeping and ensuring that all the regulatory requirements are fulfilled are very essential for a factory to ensure good quality, productivity and at the same time ensuring the employees' well-being.

It is necessary for all in the industry to plan their activities in such a way that it is safe, follow the safety regulations, provide safety gadgets, implement safe systems, periodically audit the safety systems and correct the lapses, educate others on safety issues and achieve targets without mishaps and without physical loss to the people working for achieving the targets.

Accidents not only affect the workers losing their livelihood but also employers in terms of compensation to be paid to the workers. Accidents are a significant cause of dispute between the workers and the management. The Factories Act, 1948 has laid down certain measures for the safety of the workers employed in the factories. There are a number of safety regulations developed by various countries and standard bodies. The safety of the people working on the machines has been given prime importance.

The safety regulations deal with a number of issues. While discussing on the machine lay out, it specifies the minimum distance between the two machines, the minimum distance between the wall and the machine, the minimum number of entry and exit points for a production hall depending on its size, the minimum height of the roof and so on, so that the workers can move freely, escape in case of fire accidents and avoid hitting himself to a machine part or to a wall.

The safety regulations also address the preventive actions in case of fire and the type of extinguishers to be used. While addressing the first aid boxes and safety gadgets, the safety regulations stress on the materials to be kept in a first aid box, maintenance of first aid box, verifying the expiry dates of the medicines, providing rest room (care taker room) in case of accidents or illness and providing ambulance facilities.

Safety measures result in improving the conditions under which the workers are employed and work. It improves not only their physical efficiency but also provides protection to their life and limb. People can work with a calm mind where adequate safety precautions are taken and the welfare of employees is taken care of. When people work with calm mind, they can concentrate on the quality and productivity.

It is suggested to prepare MSDS for all the products and educate the customers regarding the safety precautions to be taken.

Providing suitable MSDS for the products helps the customers in preventing damages and losses and making the best use of the materials. Now many governments are making it mandatory that MSDS is issued along with the materials wherever chemicals are used in process or as raw materials. The auditors verifying the compliance to QMS, EMS and or Legal and Regulatory requirements verify the MSDS received and also the precautions taken for the safety of employees and customers using the materials.

The following is an example of Material Safety Data provided by Akra Polyester, Mexico. Depending on the requirements of the customer, the severity of losses in case of improper use and the statutory requirements specified by the nations, the format can be modified.

MSDS MAN-MADE FIBRES POLYESTER FILAMENT TEXTURISED YARN

Party – Akra Polyester

Section I: General Data

Elaboration date:	17/08/2009
Actualisation date:	27/07/2010
Elaborated by:	Akra Polyester S.A. de C.V. Manufacturer: Akra Polyester S.A. de C.V. Ave. Ruiz Cortines y Privada Roble Nr. 100 Monterrey, NL. CP 64400 México Phone number: Tel: +52 (81) 88653100

Section II: Substance Data

Chemical name or code:	Polyethylene terephthalate with fibre finish
Commercial name:	Texturised polyester filament/DTY = draw texturised yarn
Chemical nature:	Polyester
Physical nature:	Filament continuous yarn, different dtex, counts and types, raw white or black spun dyed
Synonyms:	Polyester, filament yarn
Presentation:	Filament yarn, bobbin, warp beam
Preparation on the:	Normally, the product is coated with a spin-finish. Typical amounts vary from 0.3% to 3.0% depending on yarn type and dtex. Typical spin-finish composition: lubricants, emulsifiers, anti-electrostatic agents. White raw yarn contains titanium dioxide 0 – 1%–5%. Black yarn is produced with carbon black addition up to 2.5%.

Section III: Substance Identification

III.1 Identification

CAS Number:	25038-59-9
UN Number:	Not available
Exposure limits:	None
Immediately dangerous to life or health:	None

III.2 Hazards identification (estimated)

Health:	1
Flammability:	0
Reactivity:	0
Special notice:	None

III.3 Hazardous components

Heavy metal content:

Substance	CAS #	Max (mg/kg)	Health	Flammability	Reactivity
Antimony trioxide	1309-64-4	233.424	2	1	0
Lead	7439-92-1	0.086	2	1	0
Arsenic	7440-38-2	0.075	3	1	0

To the best of our knowledge, polyester yarn products have not been associated with negative effects on humans if used in accordance with good industrial hygiene practice and any legal requirements

Contd...

Contd...

Section IV: Physical and Chemical Properties	
Boiling point:	Not applicable
Melting range [°C]:	250–260
Flash point [°C]:	350–400
Ignition temperature [°C]:	about 420
Density [g/cm3]:	1.38–1.42
pH:	Not applicable
Molecular weight:	Not available
Physical state:	Solid
Colour:	Raw white or black spun dyed
Odor:	None
Evaporation rate:	Not applicable
Solubility in water:	Yarn product is not water soluble
Vapor pressure:	Not applicable
Percent volatile:	Not applicable
Explosive limits:	Not explosive
Form:	Multifilament yarns, wound up on suitable spools or cylinders
Thermal decomposition [°C]:	Starts at about 350°C. At 150°C evaporation of spin-finish may start.
Section V: Fire and Explosion Data	
V.1 Fire-fighting measures:	
Fire extinguishing agents:	All usual extinguishing media may be used.
Personal protection:	According to good practice of textile industry.
Restrictions:	No restriction-specific information for fire-fight: Use autonomous respirators to fight fire indoor or in poorly ventilated areas. Do not use water as extinguishable agent in the presence of uninterrupted electrical installations.
Hazardous decomposition products:	Depending on temperature and oxygen availability, combustion of gases contains variable quantities of toxic substances (carbon monoxide, acetaldehyde, various organic compounds which must not be inhaled).
Section VI: Reactivity Data	
Stability:	Stable under normal storage conditions
Incompatibility:	None known
Combustion off-gases:	Carbon monoxide, acetaldehyde, various organic compounds
Spontaneous polymerisation:	Will not occur
Conditions to avoid:	None
Section VII: Health Hazards and First Aid Measures	

	Hazard	First aid	Precautions
Inhalation	None*	No special treatment necessary	None
Eyes	Irritation	Immediately flush eyes with plenty of water. Get medical attention if irritation persists.	Wear safety goggles
Skin	May cause irritation after long-term contact	Wash with soap and water. Get medical attention if irritation persists.	None
Ingestion	None	No specific cautions necessary. Get medical attention if required.	None

Contd...

Contd...

*Fibre fly, dust and decomposition products of the finishes are to be avoided by means of removal by suction and ventilation. If exposed to excessive levels of fibre dust or fly, remove to fresh air and get medical attention if cough or other symptoms develop.

Carcinogen:	No
Mutagenic:	No
Teratogenic:	No
Median lethal concentration:	Not available
Median lethal dose:	Not available
Special precautions:	None
Other health hazards:	None
Antidotes:	None

Section VIII: Accidental Release Measures

In case of accidental dispersion or spill, collect material for controlled re-use or waste disposal. No special safety risks to environment.

Section: IX: Special Protection for Emergency Situations

No special risks if handled in accordance with good industrial hygiene practice and any legal requirements.

Section X: Transport Information

There are no data available. Not classified as hazardous.

Section XI: Ecological Information

The texturised yarn product is not associated with ecological problems, provided that the wastes are orderly disposed off.

Due to its chemical nature, the product is not eco-toxic and not readily biodegradable.

Section XIII: Handling and Storage

Fire precautions:	Remove dust, fly and finish residues by ventilation and vacuum cleaning, especially on heat-setting operations. Keep away from ignition sources. Beware of static electricity and discharges.
Storage:	All bobbins, storage cartons and bales to be stored in accordance with relevant provisions and good material handling practices.

Section XIV: Toxicological Information

The yarn product does not present special risks on humans if used in accordance with good industrial hygiene practice and the concerning legal requirements. Spin-finish on the yarn may cause skin irritation at long-term contact.

Section XV: Disposal Considerations

Can be disposed of as solid waste or burned in suitable installations, subject to local regulations. Recycling or thermal recycling is recommended for both the product and packaging material.

Section XVI: Regulatory Information

The fibre product is not a hazardous product in the meaning of national and international regulations/provisions.

Section XVII: Other Information

The information contained in this safety data sheet for man-made fibres refers exclusively to fibre product described herein. It covers neither its use in combination with any other material/preparation/product nor its use in any process. The objective of the safety data sheet for man-made fibres is to provide adequate information to commercial uses of man-made fibres and thus to make a contribution to the protection of human beings and the environment. It is not intended for the private final consumer. The information contained in this safety data sheet for man-made fibres reflects the standard of knowledge of the party completing the sheet at the time of printing. It is no contractual warranty of properties.

5.2 Work Quality Management

WQM is complementary to the work management explained earlier. One can invest and start an industry, procure right raw materials, employ highly qualified and skilled employees, provide all infrastructure supports, but still cannot ensure that he gets the required quality and productivity all the time at the specified cost and delivery in time as demanded by the customer. It is easy to prepare plans and schedules, procedures and work instructions, process flow charts, process parameters, control and check points, but it is not easy to make everyone work by heart and maintain the same morale all the time. It is necessary to adhere to the concepts of WQM to get the best out of the situation.

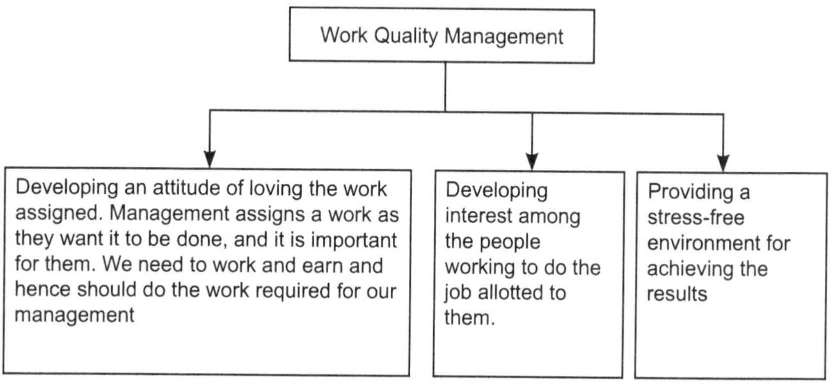

Fig. 5.2: Components of Work Quality Management

By managing work quality, with the same machine and same setting, the mistakes are not there as the work is done by heart. Only good quality is produced. The quality of the work is different from the quality of the product. When a person does a work with interest and ensures that the product meets all the requirements of the customer, the quality of the work is bound to be good. A person with interest voluntarily verifies the quality and ensures it as meeting customer requirements instead of just adhering to the process sheet given to him. He will not wait till the quality inspector comes and verifies the quality. It requires support from the top management to keep the employees motivated in achieving the work quality by the fool proof systems. The systems developed should ensure that the stress is reduced at work. A good system designed with judicial thinking and implemented with proper empowerment and whole hearted involvement of people can lead to achieve good work quality.

Many employees these days feel they are working harder, faster and longer hours than ever before and often complain of stress at work. Job-related employee stress can lead to lack of commitment to the company, poor productivity and even leaving the company, all of which are of serious concern to the management.

The product quality is limited to that product which we are checking and or offering to a customer. To get that quality, we need to have quality in the raw materials, quality in the machines that are working, quality in the people handling the operations, quality in the technology adapted and quality in the systems adapted. Good quality raw materials can be procured by paying higher price, good machines can be procured by spending more, good people can be attracted by higher salary, good technology can be identified by taking help of the experts in the field, but good systems cannot be borrowed. One can refer to a good system working elsewhere but cannot implement unless he understands, believes and adapts it as a culture by heart. Good systems are to be developed in the organisation by team work of the management and employees and should be sustained by respecting the system by the top management. If the top management respects the systems developed, the other employees are bound to respect and follow it.

The objective of WQM is to achieve the required Company Objectives while keeping the people involved including the stake holders happy at all levels. The achievement of work quality can be measured by various ways as listed below.

(a) Delighted customers because of timely supply of quality goods and services at appropriate time and all the time.
(b) Delighted customers because of prompt and quick response to their enquiries.
(c) Happy employees with no attrition, no absenteeism and no grievances.
(d) Employees are willingly taking higher responsibilities and referring their friends and family members to join the company.
(e) No expenses relating to fighting legal cases in the court and no notices received by any of the legal or statutory bodies for not meeting the compliance to any of the legal requirements.
(f) No loss of production due to any reason such as breakdowns, lack of program, poor working, short of materials, lack of clarity in instructions and so on.
(g) Reduced cost of manufacturing.
(h) Increased sales turnover per employee.
(i) Increased sales turnover per rupee spent in marketing.

(j) Reduced rejections.
(k) No delay in delivery of any order.
(l) No material returned back from market for any reason.
(m) No complaint or grumbling (bad reputation) from market not only regarding the product quality but also due to other reasons such as poor response, not responding properly, not treating a customer properly, delays in transportation, poor transporter, discrepancies in sales documents, discrepancies in markings, shortage or excess supply, price charged higher than market rate, and so on.
(n) Zero accidents.
(o) Zero fires.
(p) Zero overtime.
(q) Highly proactive in getting information from the market and changing their strategies and products to meet the changing requirement and never blame recession for the losses if any.
(r) Always being on toe to face competition and continuously stretch their targets after accomplishment of each target.
(s) Being a benchmark for others and so on.

The employees should be clear about the purpose for which the texturised yarn is being produced and the critical points that can hamper the performance at the customer's end. This helps the operatives to concentrate on those parameters that are critical and help achieving customer satisfaction.

The quality and efficiency can be increased by proper coordination of works of different people and all members contributing equally in achieving the targets with interest and commitment. The work quality of the workers, supervisor, the manager and the management are all important in achieving the quality and efficiency in texturising.

To do any work, we need to develop, establish and define a procedure and work accordingly. If the procedures are not defined, there are chances of failures. The majority of the problems we get in the industry are for not defining the procedures and/or not following the defined procedures. People want quick results and hence take short cuts and finally end up in a mess. The main culprit is the top management, who even after knowing that the procedure has to be followed insist the juniors to bypass the procedure and do the work. Bypassing the authorities, avoiding quality inspection, not analyzing the vendor capabilities before purchasing materials, not verifying the minimum competency level while recruiting people for job, not recording the extra hours worked and not paying the workers as per the law, promoting a person without verifying the competencies and neglecting other people in the organisation eligible for promotion, not installing safety systems and not checking safety systems and so on are some examples. All these leads to poor work quality.

5.3. Total Quality Management

Achieving the required quality while adhering to delivery schedules along with maintaining the cost at lower level all the time should become the culture of an organisation to ensure successful working. Short-term approaches and shortcuts may be felt as profitable or smartness, but can kill the organisation in the long run. When people working in all sections of an organisation involve themselves in producing the right quality in all the processes and at all the time, the product produced is bound to be good. TQM is limited not only to the product quality but also to all the aspects of running an organisation such as products, services, systems, culture and ethics.

Fig. 5.3: Quality at all places

Total quality means quality at all levels. No work is considered as inferior. All activities are important in an organisation and all have to be done as per their expected norms at all the time. This culture helps in achieving the quality all the time. To achieve this, all have to work as one team and the top management should be a role model. People working should be clear in their roles and responsibilities and should have required empowerment to do the activities. It includes working out the procedures for doing the work, involving the people to do the work, educating, training and motivating the people, doing the work, reviewing the works, correcting the mistakes and taking suitable preventive actions. It is an organisation wide approach towards quality which results in performance excellence in meeting customers as well as stakeholder's requirements by doing the right things right at first time and every time.

The following are the imperatives for successful implementation of TQM concepts:

- Management leadership
- Product, process and system excellence
- Human resource excellence
- Continuous improvement
- Customer orientation
- Respect to the community
- Respect to the legal and regulatory requirements
- Sincere approach to meet the requirements of stake holders
- Attitude to accept and correct mistakes

The basic steps in implementing TQM are as follows:

1. Top management education
2. Quality policy
3. Setting up of TQM cell
4. Organising quality councils
5. Training of trainers and facilitators
6. Middle management education
7. Supervisory staff education
8. Workers' education
9. Setting up of cross-functional teams
10. Creating climate for quality circle activities
11. Setting up of quality circles
12. Writing of standard operative instructions/procedures

13. Training of internal quality auditors
14. Internal audits
15. External audits
16. Review process of cross-functional teams
17. Review progress of quality circles
18. Re-education – success experience

Defining objectives is the main responsibility of the top management, specifying clearly the production, quality, cost and delivery requirements. The top management fixes the objectives for the company, departmental heads fix the objectives for their departments, sectional heads fix objectives for their sections and finally individual targets are set. The top management works as a role model and communicates the importance of meeting the customer requirements as well as the regulatory requirements by involving and empowering the people to work willingly for the improvement of the systems and the organisation.

For ensuring involvement of people, the following steps are essential:

- Explain the importance
- Explain the roles
- Define the authorities and responsibilities
- Explain the system
- Clarify the doubts
- Help in understanding and performing
- Recognise the work
- Give credit to teams

Whenever a deviation is noticed, a corrective action is to be taken. A very important aspect is to identify the root cause for that deviation and taking preventive actions. There are different problem-solving techniques and the management should identify the correct technique and use it.

5.4 Housekeeping and Material Handling

When we analyse the reasons for poor quality, the habits and culture of an organisation are found having maximum share. The present-day modern machines are so well designed that the parameters once set shall not deviate unless someone meddles with it. The major reasons for poor quality are poor housekeeping, dust or oil contamination and poor handling of the produced materials leading to coil slippage or damaged packages, which are all avoidable provided the people and the management are strict in maintaining proper housekeeping and adopt proper system of material handling. The synthetic filaments are more vulnerable to handling damages compared with the spun yarns.

Practicing the 5-S principles help a lot in maintaining the quality of the products. They are as follows:

- ***Seiri*** → Sorting out the unwanted materials
- ***Seiton*** → Systematic arrangement of the required materials
- ***Seiso*** → Spic and span cleanliness and inspection
- ***Seiketsu*** → Standardising
- ***Shitsuke*** → Self-discipline and training

Sorting out the not required materials and removing them from the place of work is the first principle of 5-S. This helps in providing place to keep the required materials and to do the activities without hurdles. The unwanted materials, unwanted processes and unwanted persons create irritation. This reduces the quality of the work and spoils the peace of mind. The concept of Seiri says that you remove the scrap away and get space released for other materials. Move items useless to you but likely useful to others to a common place which shall help in eliminating the unwanted purchases.

Seiton means systematic arrangement of the materials. We need to arrange them in such a way that we get them when needed. No one should waste time or energy for searching. Store away the materials used less frequently and retain the materials frequently used near the workplace. This will avoid confusions and mix ups and improves the efficiency. Preparation of a time log and recording the activities and grading them as per their importance can help in prioritising the activities.

Seiso talks of spic and span cleaning. This concept includes the cleaning of everything at workplace and inspection for abnormalities during cleaning and routine maintenance, thus reducing the chances of breakdowns. A clean work area is pleasant for working and work efficiency and the quality shall improve. An unclean and congested area reduces the work efficiency. When the environment is clean, man can think and find solutions to his problems. If the area is clean, the defects shall be visible.

Aisles and working spaces should be kept in good order without any obstacles and space for free movement of men and materials. A congested alley leads to a number of accidents such as the workers cloth or body touching the running parts of the machine, putting step on slippery materials, hitting another person coming in the opposite direction and so on.

The material handling systems should ensure that no filaments or yarns are touched by hand or any other yarn packages or mechanism. The holding of the packages should be only at the paper tubes. The stacking of the packages should ensure that no load is exerted on the yarn packages and they are stacked in a clean dust-free environment.

Fig. 5.4: Improper storing of yarns

Improper storing of texturised yarns, which leads to package damages and also to more wastage. It creates more breakages in further process and also can result in stains.

Fig. 5.5: Broken crates

Using of broken crates damages the yarn

Fig. 5.6: Pegged trolleys

Pegged trolleys are suggested for transporting packages within a factory

Fig. 5.7: Packing each package

Each package should be covered properly and kept in carton with separators at all sides

Fig. 5.8: Source of mix up

One package of different lots is kept with this – this leads to shortage of material in critical times

Fig. 5.9: Packing of POY

The proper packing of the materials adds value to it and attracts the customers. If packing is not proper, although the basic quality of the material is good, it shall not be accepted by any customers.

5.5 Training and Education of People on Work

The success of a company depends on how competitively one can produce goods to the satisfaction of the customers at competitive price and deliver in time. Therefore, continuous developments can be seen in the technological aspects as well as in managerial aspects in the industry.

Knowledge is very important for the success of a system in any industry. As the customer needs are changing and progressing, the knowledge essential for the progress needs to be shared with the people considered. The organisation has to ensure that the required knowledge is provided to their staff so that they can perform. The knowledge about the raw materials, knowledge about the maintenance of machinery, knowledge about the procedures to be adopted, knowledge about the testing to be done, knowledge about the areas of concern and knowledge about the corrective and preventive actions to be taken are very important. The knowledge shall be shared with the people may be by discussions or by providing documented information such as procedures and instructions. Wherever practical, demonstrations are to be made and practical exercises are to be conducted to develop skill among the people to face the situation and handle the machines and materials as needed. The knowledge has to be protected so that the competitors do not get the secrets of the processes; in the meantime, it should be ensured that people working on the shop floor get the required knowledge for doing the work.

Depending on the requirements of individual mills or the factory, different modules can be developed for training of staff and workmen. The modules should be prepared considering the specific works to be done and the existing level of competency of the people on the job. The course curricula should contain the practical competencies, the underpinning knowledge to attain that practical competency, the tools and equipment needed, the control points and the check points in each operation, the effect of various aspects on the quality and production, the critical quality requirements and the soft skills for the supervisors to manage their sections.

The education and training given needs to be specific to each employee depending on the activities assigned to him. One should be clear on the jobs expected from an employee and the authorities and responsibilities. Therefore, it is necessary to list and document the job descriptions and train the people to perform those jobs and develop skills to perform them effectively.

An example of job descriptions of an operator tending a yarn texturising machine is given below.

5.5.1 Guidelines for Yarn-Texturing-Machine Operator

Jobs done

- Attends texturising machine equipped with thermal units and false-twist spindles to produce stretch yarn.
- Threads the yarn through guides, feed rollers and heating unit, around the entrance wheel of false-twist spindle, through traverse guide, and onto take-up bobbin, utilising suction lines along the yarn path, extension rods and wire threader.
- Observes signal lights that indicate the yarn breaks and malfunction of specified sections of the machine.
- Attends the breakages.
- Records information regarding malfunction on flag card and places card on the machine unit to signal maintenance crew.
- Reads doff schedule or uses gauge to determine when the packages should be doffed.
- Doffs full take-up packages.

Responsibilities

- Maintaining the machine and its surroundings always clean.
- Checking the labels of the POY packages for lot number, denier and so on before creeling them on the texturising machine.
- Cleaning the yarn path before inserting a new yarn in the machine.
- Keeping a vigil on the temperature of the heat-setting units and controlling it within the limits instructed.
- Attending the yarn breaks and run outs in time.
- Turning out the required production as per the speed of the machine for the specific twist and denier.
- Doffing the correct size packages.
- Informing the mechanic/supervisor in time regarding any problems faced.
- Handling the materials properly without causing damages due to handling.
- Pasting stickers to the produced materials as instructed by the superiors.

- Keeping aside the rejected materials.
- Not allowing any unauthorised person to touch the machine or materials.
- Cleaning his hands and ensuring it as dirt free before handling the materials.

Authorities

- Stopping a malfunctioning machine or production unit and informing the mechanic.
- Rejecting the feed materials in case of improper label or identification.
- Rejecting the material produced from a defective production unit.
- Asking for clarifications when the instructions are not clear.
- Stopping any unauthorised person touching the machine or materials.

Trainings needed for

- Threading the yarn as per requirements.
- Cleaning the yarn path.
- Operating the texturising machine.
- Setting and controlling the temperature and speeds.
- Visual checking of the produced texturised materials.
- Visual checking of the feeding material.
- Checking the labels and empty tubes.
- Attending to minor settings to correct the quality.
- Handling the supply packages and produced materials without damaging the layers and coils.
- Keeping his hands always clean while handling the materials.
- Reading work instructions, setting the machines as needed and working as per the instructions.

Minimum competency required

- Basic educations to identify the materials and to read the labels and work instructions.
- Good eyesight to detect the breakages and poor quality.
- Sharp in identifying the changes in machine sound and deviations from normal.
- Physically fit to work continuously on the machine for the complete shift.

5.5.2 Roles and Responsibilities of Supervisors

The supervisor is the manager of his section and has the objectives of achieving the sectional and departmental objectives by planning, coordinating, implementing and controlling the activities. The supervisors, who are the mediators between the top management and the bottom line workers, have the following responsibilities:

- Understanding precisely the requirements of the customer and communicating to the workers producing the quality in the language understandable by them.
- Understanding the company's capabilities, such as process capability, human resource capability, funds' availability, material availability and infrastructure availability and working out the plans to accomplish the production requirements.
- Understanding the ethical and legal requirements of the process for the products and services planned and monitoring to meet the requirements.
- Designing the product that is acceptable to the customer and the society.
- Designing the process to get optimum output at minimum time and cost.
- Deciding the measures for measuring and monitoring processes and monitoring accordingly.
- Working out the detailed quality plans and the actions needed at different levels and ensuring the same is implemented as planned.
- Working out the production program, implementing and monitoring them.
- Planning for the raw materials, spares and consumables and indenting for the same as needed.
- Procuring the required materials in time and ensuring their quality by proper inspection and testing.
- Planning the maintenance activities and implementing them to have minimum loss of time for maintenance while ensuring lowest possible failures.
- Tuning the machines as per the process designed while understanding the technical capabilities.
- Educating and training the men on shop floor on the production and quality aspects of the product and the process.

- Allocating suitable competent workmen for the skilled jobs and monitoring their works relating to quality and productivity.
- Maintaining harmony in the workplace and creating a good work environment.
- Monitoring the process periodically to ensure its suitability and to have lowest wastes possible.
- Documenting the procedures adopted, actions taken on problems and deviations in the results, monitoring and ensuring following of the procedures all the time.
- Reporting the activities suitably to the superiors as well as to the next person taking charge highlighting the deviations and special actions.
- Analyzing the reasons for deviations in process and product performance and discussing with the superiors for suitable actions and taking actions as decided.
- Arranging for monthly, quarterly, half yearly and annual stock taking as per the requirement of the company and the section.
- Working out the realisation of the fibres, yarns and fabrics on periodic basis referring to actual consumptions and stocks and the material actually realised.
- Maintaining harmony at work by adhering to company's policies, legal and regulatory requirements, while understanding the needs of stake holders and taking all the workers together to achieve the objectives and goals.

5.6 Maintenance Management to Ensure Quality

Good maintenance practices not only ensure low material cost, low manpower involvement, less time duration in the process and optimum utilisation of resources increase the productivity but also ensure defect-free products. The need for maintenance is predicated on the actual impending failure; ideally, maintenance is performed to keep equipment and systems running efficiently for at least designed life of the component. As such, the practical operation of a component is a time-based function. There are different practices of maintenance.

- ***Unplanned or breakdown maintenance:*** In this case, the machine is allowed to work till it stops by itself due to a breakdown. This is a bad practice as the breakdowns occur when a part is worn out and is unworkable and might damage the adjacent parts also. Before the part reaches the stage of break down, it shall be continuously wearing out and the quality shall be deteriorating. When a breakdown occurs, all the materials running on the machine get spoiled resulting in huge wastage. The breakdowns result in the stoppage of the machines, which was unplanned and affects the delivery schedules.

- ***Preventive or periodic maintenance***: In this case, a maintenance schedule is prepared depending on the type of the machine and the mechanisms to be serviced periodically. Each part is checked when the machine is stopped for maintenance and is corrected or replaced. The parts which have a predetermined working life shall be replaced during this maintenance activity. This helps in preventing the unplanned stoppages due to breakdowns.
- ***Corrective maintenance:*** In this type, actions such as repair, replacement or restore will be carried out after the occurrence of a failure to eliminate the source of this failure or reduce the frequency of its occurrence. The failure need not only be the breakdown, but may be a poor quality of the materials produced, which normally happens due to disturbed settings, vibrations or improper handling. The corrective maintenance is subdivided into three groups.
    - *Remedial maintenance*: It is a set of activities that are performed to eliminate the source of failure without interrupting the continuity of the production process. The way to carry out this type of corrective maintenance is by taking the item to be corrected out of the production line and replacing it with the reconditioned item or transferring its workload to its redundancy.
    - *Deferred maintenance*: It is a set of corrective maintenance activities that are not immediately initiated after the occurrence of a failure but are delayed in such a way that will not affect the production process.
    - *Shutdown corrective maintenance*: It is a set of corrective maintenance activities that are performed when the production line is in total stoppage situation.
- ***Predictive maintenance:*** Predictive maintenance is a set of activities that detect the changes in the physical condition of equipment (signs of failure) to carry out the appropriate maintenance work for maximising the service life of the equipment without increasing the risk of failure. It indicates the design of new equipment. The routine testing of the materials produced gives an indication of trends in deterioration of parts. The weakness of the current machines is sufficiently studied

(on-site information leading to failure prevention, easier maintenance and prevents defects, safety and ease of manufacturing) and is incorporated before commissioning a new equipment. It is classified into two kinds according to the methods of detecting the signs of failure:

(a) *Condition-based predictive maintenance*: Condition-based predictive maintenance depends on continuous or periodic condition monitoring equipment to detect the signs of failure.

(b) *Statistical-based predictive maintenance*: Statistical-based predictive maintenance depends on statistical data from the meticulous recording of the stoppages of the in-plant items and components to develop models for predicting failures.

The drawback of predictive maintenance is that it depends heavily on the information and the correct interpretation of the information. Some researchers classified predictive maintenance as a type of preventive maintenance.

The main difference between preventive maintenance and predictive maintenance is that predictive maintenance uses monitoring the condition of the machines or equipment to determine the actual mean time to failure, whereas preventive maintenance depends on industrial average life statistics.

- ***Improvement maintenance (IM):*** It aims at reducing or eliminating entirely the need for maintenance. That can be occurred to simplify maintenance work, which may increase the overall performance. This type of maintenance is subdivided into three types as follows:
  - *Design-out maintenance*: A set of activities that are used to eliminate the cause of maintenance, simplify maintenance tasks or raise machine performance from the maintenance point of view by redesigning those machines and facilities which are vulnerable to frequent occurrence of failure, and their long-term repair or replacement cost is very expensive.
  - *Engineering services*: It includes construction and construction modification, removal and installation and rearrangement of the facilities.
  - *Shutdown improvement maintenance*: It is a set of improvement maintenance activities that are performed while the production line is in a complete stoppage situation.

To achieve the required quality of the products, it is essential that the machine is always in good condition and kept clean. Periodic checking of the machine parts, replacing the worn-out or deformed parts, lubricating the bearings and preventing the parts from getting worn out are very essential.

The maintenance schedules are to be decided on the type of the machines, the materials being worked, the speed of working and the normal concerns of the customers regarding product quality requirement.

Cleaning of the yarn path especially the friction discs and inside portion of heat-setting units greatly reduces the faults in a texturised yarn. The frequency of cleaning depends on the material being worked and the deposition of spin-finish, burnt particles (fibre snow formation) and fluff. Overmaintenance, that is, opening some mechanisms frequently and reassembling might cause more harm than not opening those mechanisms. One should be clear as to which mechanisms are to be allowed to open by shop-floor people and where the help of the experts from machine manufacturers is to be taken.

At the End

The quality expectations of the customers are to be understood clearly and the yarn should be engineered to meet the customer's requirements. Proper selection of raw materials, machines, training of workmen, developing suitable operating procedures and work instructions, educating all on implementing systems, proper maintenance, periodic reviews, continuous follow-up empowering people to do their jobs, eliminating work-life stresses, developing a culture of TQM, ensuring proper handling, packing and presenting the materials and so on are all important in producing defect-free texturised yarns and winning the hearts of the customers.

6
Testing of Texturised Yarns

As a part of quality assurance, one has to test and confirm that the materials produced are suitable to the purpose of the customers and are acceptable as per the norms agreed with the customer. The normal tests conducted to find the suitability of a texturised yarn are tensile testing, denier testing, yarn skein shrinkage, degree of bulk, degree of crimp, degree of texture, stability of bulk, stability of crimp, stability of texture, assessment of expected cover in fabric, package-to-package dye uniformity and visual inspection before despatch.

The properties shown below are specific to the texturised or drawn yarns. They are generally not applicable to POY yarns.

- Tack retention: The ability of the tacks to withstand stretching. This is important when processing imparts different degrees of strain on the yarn. Tack retention quantifies the 'strength' of the tack to withstand strain and, therefore, provides bundle cohesion for processing.
- Twist direction: The rotational direction by which twist is present in a yarn. It can be designated as 'S' or 'Z'.
- Twist level: The number of turns of twist in a yarn per a given length of yarn. For the twisted yarns, the level of twist is critical for the degree of bundle cohesion needed for processing and yarn strength. In the texturising process, the twist level has a significant impact on the crimp level of the yarn and also the fabric properties as per the type of the fabric.
- Torque: The tendency of a yarn to rotate and kink upon itself in an effort to relieve twist. The impact of torque, causing kinks in the yarn, can lead to processing problems. Torque can often be observed in the fabrics. Torque can be balanced by plying 'S' and 'Z' yarns, or minimised with jets.
- Skein shrinkage: The length contraction of a skein of yarn, held at a prescribed load and after heating under prescribed conditions. Tests are designed to simulate various down-stream processing conditions. Some tests simulate the restraining forces experienced in fabric dyeing and finishing, and are used to design the texturised yarns for different types of fabric applications.

- Residual skein shrinkage: The residual shrinkage test utilises a much lower restraining load than other tests and is a better indicator of the ultimate shrinkage potential of the yarn

6.1 Measurement of Crimp Properties

The traditional test for measuring crimp and shrinkage is an effective way to characterise crimped filament yarns. In this case, the length of a crimped yarn is measured without stretch and the length after stretching. The percentage crimp is calculated by dividing the difference in length by the length of the yarn with crimp. However, it has several limitations in its application such as the slowness of the test, the limited statistical representation of the product and the human error.

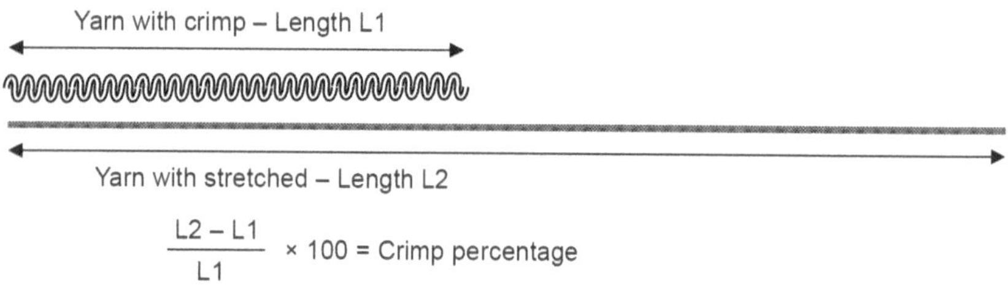

Fig. 6.1: Measuring crimp – Conventional method

ASTM D6774 – 02 (2010) Standard specifies a test method for crimp and shrinkage properties for the texturised yarns using a dynamic texturised yarn tester. For some yarns, the elapsed time between processing and testing has a marked effect on the results, of this test, especially during the first 72 hours. The effect is caused by stress decay which is known to be minimal beyond the seventh day and after which time the yarn remains relatively stable. Therefore, specimens should only be compared if tested after the same elapsed time. Samples can be tested at-line, thus having little to no elapsed time between processing and testing. This test method covers the determination of crimp contraction, residual fibre shrinkage and their variability of all types of filament yarns (POY, FOY, flat yarns, texturised and bulked continuous filament (BCF) carpet yarns) using an automated tester. This test method is limited to crimped, multi-filament yarns ranging from 22.0 to 890 dtex (15–800 denier) and for BCF yarns from 890 to 4200 dtex (800–3800 denier).

Crimp frequency, amplitude, crimp stability, crimp elongation and de-crimping point are some of the important properties that determine crimp. Crimp frequency and amplitude may be determined by projecting a magnified image of the fibre on screen. Optoelectronic sensor is used in some equipments to provide digital representation of the fibre held between the two clamps at a very low tension. Tensile tests using an extremely sensitive force-measuring system enable the measurement of the curve of crimp force versus elongation, crimp extension, de-crimping point and crimp stability.

Crimp is defined as the waviness of a fibre, yarn or tow. This characteristic may be expressed numerically by the combination of the crimp frequency either with the crimp contraction or, in the case of the texturised yarns, with the crimp elongation.

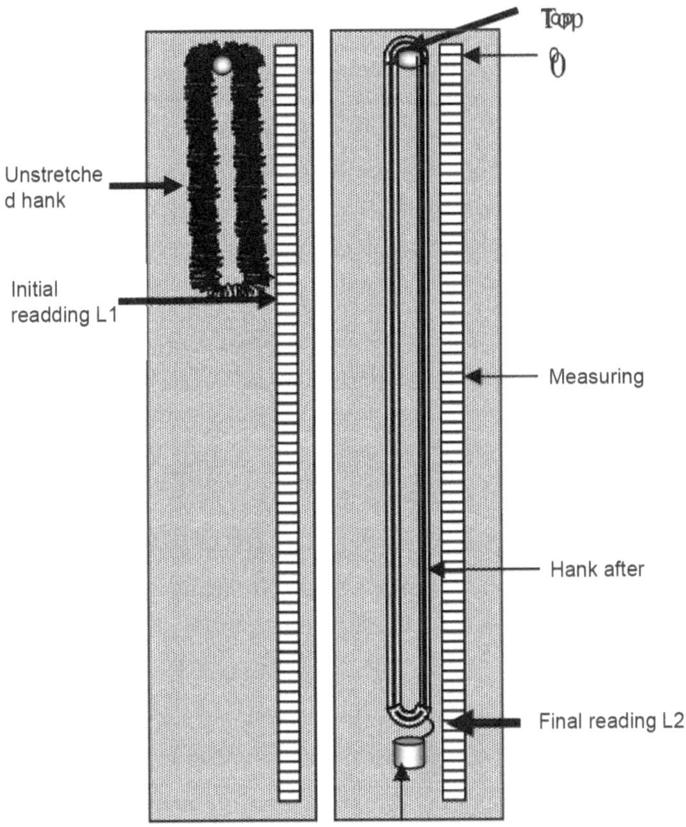

Fig. 6.2: Measuring Crimp by stretching skein of texturised yarn

- **Crimp contraction or percentage crimp** is the contraction of a crimped fibre or a texturised yarn owing to the development of crimp, expressed as a percentage of its straightened length.

- **Crimp elongation:** The lengthening of a crimped fibre or of a texturised yarn after development of a crimp when it is straightened under specified tension expressed as a percentage of its initial length.
- **Crimp frequency:** The number of crimps per unit length of the filament yarn. Different methods are in common use for expressing crimp frequency, based on half or whole waves and on straightened or unstraightened length. These bases must therefore be specified in any quantification.
- **Crimp, latent:** Crimp that can be developed by a thermal treatment or by tensioning and subsequent relaxation.
- **Crimp liveliness:** The tendency for a texturised yarn to develop its crimp immediately after the reduction in an applied tension.
- **Crimp stability:** The ratio of the crimp of a fibre or texturised yarn after a specified treatment to the crimp prior to treatment, expressed as a percentage. The method for determination of the crimp and the treatment must be reported.

The most important quality problems of BCF yarns are the differences in crimp and shrinkage properties leading to unevenness of the carpet both with regard to yarn bulk and dyeability (for polyamide yarns). For many years, a typical laboratory set-up for a texturised yarn producer included the traditional skein test for measuring crimp contraction and shrinkage properties because of the strong effect they have on the dye-ability of the yarns. For this reason, dynamic testing of crimp, shrinkage and entanglement has emerged as a preferred method by many companies compared with the traditional method used.

A simple method of testing the crimp is shown in Figure 6.2. A skein of the texturised yarn is hung on a support vertically on a board, which had a graduate scale fitted. The initial reading L1 is noted down without stretching the skein. Then a fixed dead weight is hanged to the skein so that the crimps become straight. The dead weight selected should be such that it straightens the yarn but does not stretch. The length L2 is noted down of the straight skein. The crimp percentage can be calculated by using the formula $[(L2 - L1)/L1] \times 100$.

Crimp frequency, amplitude, crimp stability, crimp elongation, de-crimping point are some of the important properties that determine crimp. Crimp rigidity tester is used to determine the bulking potential of the texturised yarns. Crimp frequency and amplitude may be determined by projecting a magnified image of the fibre on screen. Tension device is used in the equipment to provide proper tension between the two clamps. A load equivalent to 0.09 g/dtex is suspended from the end of a yarn immersed in water. After a fixed time, the load is reduced to 0.0018 g/dtex, and after a further fixed time the reduction in the length of the yarn is measured. The crimp rigidity of the yarn is the reduction in the length of the yarn expressed as a percentage of the original length.

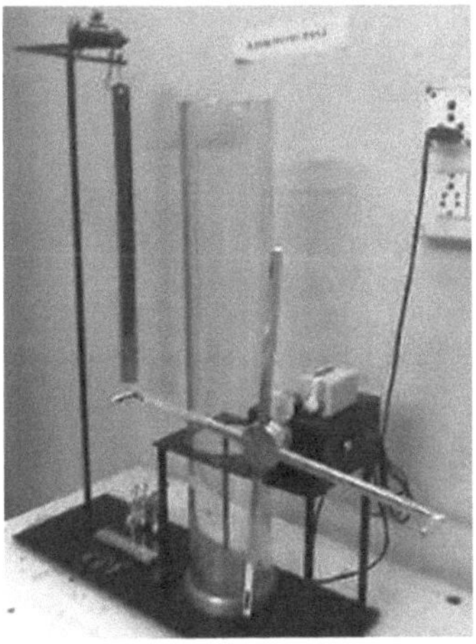

Fig. 6.3: Crimp Rigidity Tester by B. Tex Testing

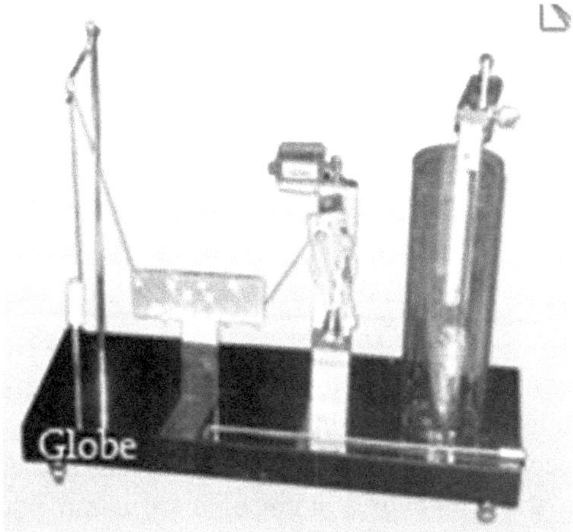

Fig. 6.4: Crimp rigidity tester by Globe enterprises

BTex testing equipment is shown in Figure 6.3. The following is the test procedure:

1. Place the package of the yarn to be tested on the stud of the base plate of the hank winder in the apparatus below the yarn guide.
2. Thread the yarn through the gate tensioner and adjust the tension until, on pulling the yarn through the gate; it can be observed that the crimp is getting removed.
3. Clamp the leading end of the yarn to the inner side of the capstan. Wind a hank of the yarn with a number of strands chosen so that loads of 0.09 and 0.0018 g/dtex can be applied to each strand by means of the sets of hooked weights provided. (The loads and number of strands for yarns most commonly encountered are listed in the table given in BIS Handbook.) Clamp the final strand to the outside of the capstan.
4. Add appropriate lightweight to the strands extending across the prongs of the fork at the lower end of the arm of the hank winder. Add appropriate heavy weight to the lower hook of the lightweight.

5. Remove the hank and the suspended weight by unscrewing the hank bar from the arm of the winder.
6. Attach the extra clip support to the hank bar and immerse the hank in the water contained in the measuring cylinder by clipping the two supports over the rim of the cylinder.
7. Start the stopwatch.
8. Just prior to 2 min from immersing the hank, measure the length of the hank as L0.
9. At 2 min from the time of immersion, remove the heavy weight by raising the grid, causing the weight to be released from S-shaped hook.
10. After a further 2 min, measure the length of the hank as Ls.

    Crimp rigidity = L0 – Ls/L0 × 100

The Globe Enterprises explain the method of testing their apparatus as follows. Crimp rigidity of a texturised yarn gives a measure of its ability to recover from stretch. It is determined by measuring the percentage reduction in the length of a skein of a yarn immersed in water when a minor load replaces a major load. For measuring crimp rigidity, a skein of known number of strands is prepared from the package of the yarn under test. This skein is then immersed in the water filled in a glass jar with a load of 0.1 g/denier/strand suspended on it. The length of the skein is measured on a scale. The load is then reduced to 0.002 g/denier/strand by replacing the major load by the minor load and the reading on the scale is observed. Crimp rigidity is calculated from the readings obtained under the major and minor loads

The bulk properties are very important to the texturised yarns, and hence testing of the same is very important. ASTM D4031 specifies the methods for testing the bulk properties of a texturised yarn. This test method covers the measurement of the change in length of a tensioned skein of the texturised yarn due to the change in crimp characteristics brought about by exposure to wet or dry heat. The change in length, depending on the procedure, is a measure of skein shrinkage, crimp contraction, bulk shrinkage or crimp recovery. This test method applies to crimped, continuous multifilament yarns ranging from 1.7 to 88.9 tex (15–800 denier). Three conditions are provided for crimp development mediums, and loading routines are provided to be used on the yarn skeins to allow determination of the yarn bulk by several different procedures.

Today, many of the texturising machines are equipped with tension control on-line. This is a very helpful addition for process control for detecting intermittent faults throughout the package; however, it offers limited dyeability assurance, as it is not very sensitive to heat-related faults (compared with measuring crimp) and will not detect tension-related faults after the measurement point. The knit-dye-grade (KDG) method is commonly used for inspection, which is slow, laborious, highly subjective and offers little quality control information.

Texturmat ME adopts the crimp contraction test procedure according to German standard DIN 53 840. This test procedure has been designed for yarns up to 500 dtex. It uses yarn hanks with an overall count of approximately 2500 dtex, the hanks being subjected to various loads during testing and their length measured at each stage. Crimp contraction, crimp modulus and crimp stability are calculated from the measured lengths 1g, 1f, 1z and 1b as shown in Figure 8.6 and the table using the following formulas:

Crimp contraction = [(lg – lz)/lg] × 100 (%)

Crimp modulus = [(lg – lf)/lg] × 100 (%)

Crimp stability = [(lg – lb)/(lg – lz)] × 100 (%)

For optimum development of the yarn crimp, the yarn hanks are subjected to a thermal treatment prior to test start.

The crimp contraction testing of the texturised carpet yarns (BCF), which characteristically have relatively high yarn counts, is a special case. For this application, yarn hanks with only one wrap, that is, a yarn loop, can be used for testing.

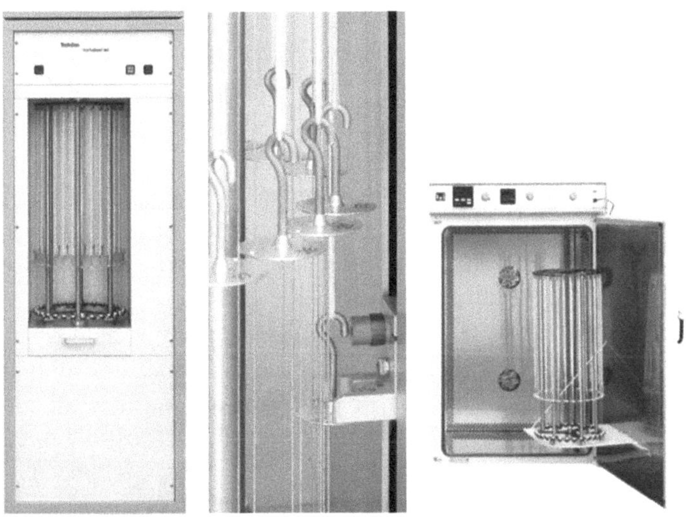

Texturmat me Pre-tensioning weights Heating cabinet

Fig. 6.5: Texturamat ME

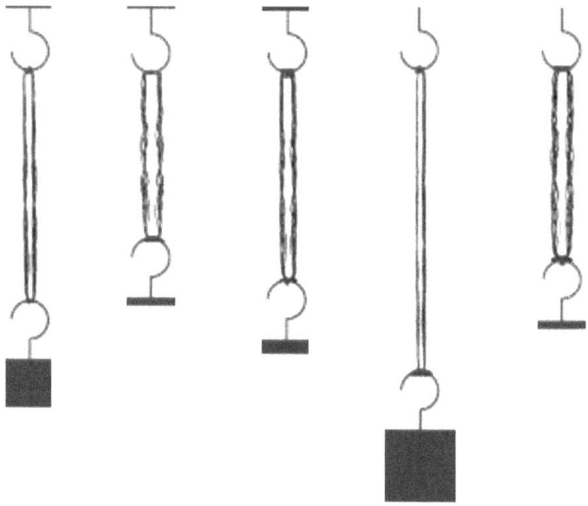

Load (cN/tex)	2	0.01	0.1	10 or 20	0,.1
Duration of load	10 s	10 min	10 s	10 s	10 min
Length of hank	lg	lz	lf		lb

Fig. 6.6: Schematic representation of the crimp contraction test

The determination of the yarn shrinkage, due to thermal or hydrothermal treatment, is a further test method for the texturised yarns and other yarn types, for example, according to DIN 53 866 or ASTM D 2259. In this instance, testing is carried out on the yarn hanks by determining an initial length (length lg1), followed by a thermal treatment and a second length measurement (lg2). Shrinkage is calculated by applying the following formula:

$$\text{Shrinkage} = [(\text{lg1} - \text{lg2})/\text{lg1}] \times 100\ (\%)$$

The texturising quality of the air-texturised yarns is determined by carrying out tests on the yarn hanks, and the proportional length increase in the hank after being subjected to a high-tensile force is the measured variable and criterion for the stability of the yarn structure.

The Texturmat ME is a test instrument for fully automatic length measurements on the yarn hanks, characterised by its high flexibility in terms of test sequence configuration, choice of measuring loads and time intervals for the loading and unloading periods. As a result, it is possible to carry out different testing methods according to different standards, for example, crimp contraction or shrinkage tests not only on (texturised) filament yarns but also on fibre tow. The instrument is equipped with a removable 30-position magazine, within which the suspended yarn hanks are pre-loaded with weights.

Special magazines are also available for testing BCF yarns. The application of tensile forces and the measurement of yarn lengths for predetermined load times are program-controlled and are brought about by a loading carriage in combination with a high-resolution force-measuring system. The size of the force can be freely selected.

The measured values, for example, crimp contraction, crimp modulus and crimp stability according to DIN 53840, are automatically determined and are evaluated by the test control PC system.

The instrument is operated with several magazines. In addition to the magazine already in use in the test instrument, others can be prepared, that is, loaded with hanks or heat-treated to ensure a continuous test cycle. A heating cabinet with the appropriate inner dimensions to accommodate the magazine loaded with the hanks is available as a Texturmat ME accessory. The control range of the heating cabinet enables a full range of temperature settings necessary for crimp development and the inducement of shrinkage.

For hot-air shrinkage tests with relatively high temperatures, it is essential that after opening the heating cabinet and inserting the magazine, the selected temperature is achieved within 30 s, as specified in the DIN 53866 standard. For this application, a special version of the heating cabinet with a computer-controlled heating system is available.

When crimp development or the inducement of shrinkage in hot water is required instead of hot-air treatment, a hot water container can be fitted that can also accommodate one magazine. The container is equipped with a jib crane, incorporating an electrically operated chain hoist, with which the magazine can be lowered and raised in and out of the water bath.

The TYT-EW from Lawson-Hemphill employs dynamic testing ASTM standard has included this method for texturised yarn titled 'Standard Test Method for Texturised Yarns Using a Dynamic Texturised Yarn Tester'; the ASTM designated number is D 6774-02.

The TYT has a two-zone measuring system, which separates the true crimp and true shrinkage from the total contraction. The TYT has many similarities to the traditional skein test except that the test is performed dynamically. It allows for limits to be set for the properties measured allowing for the faulty position on the machine to be quickly identified. The data can be automatically transferred via an Ethernet connection to a LAN system through Windows networking protocol. Factories can use the TYT effectively as a process control tool for evaluating quality, fixing problems in production quicker, improving quality, provide better analysis of the true production situation and for R&D with new products. The TYT-EW is claimed to be very effective in high-speed testing up to 1000 packages in 8 hours. It is provided with SAK industrial package changer that allows up to 36 ends to be preloaded from a creel directly.

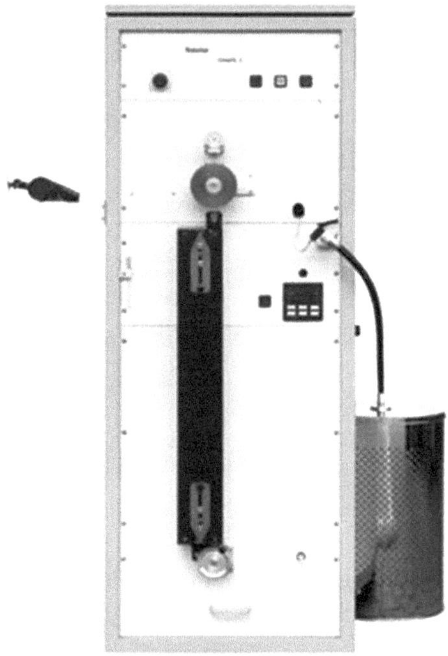

Fig. 6.7: DYNAFIL C

Dynafil M and Dynafil C are developed by Textechno for testing the crimp force in the texturised yarns. The Dynafil C incorporates the two-stepped godet technology, enabling arbitrary extension settings between –50% and +400% through exchanging the godet rolls. For the most important application areas of crimp-force testing on the texturised yarns, draw-force measurement on pre-oriented yarns and shrinkage-force examinations, only few length variation steps are usually required, that is, –5% (texturised yarn), 0% (flat yarn) and 60% (polyester POY) or 30% (nylon POY).

Further important technical characteristics of the Dynafil C are as follows:

- Godet drive up to 500 m/min
- Convection heater with a temperature range up to 300°C
- Load cell with measuring ranges 20, 100 and 1000 cN
- Feed wheel pre-tensioning device

The equipment is equipped with a PC system for controlling the test processes and for the evaluation of the measured data. It is also possible to add the automatic package changers Type SE or SM or to carry out friction measurements.

Dynafil ME: The newest evolutionary stage in the Textechno Dynafil series is the Dynafil ME. Apart from a substantial increase in the output, the Dynafil ME is characterised by a high degree of automation which is user-friendly.

The earlier Dynafil models worked with constant overfeed/extension. In addition to this, Dynafil ME enables measurement of extension or contraction at constant tensile force on the running yarn, for example, percentage crimp contraction of the texturised yarns. Furthermore, it is possible to carry out time-dependent shrinkage and shrinkage-force measurements and relaxation or creep tests on the stationary yarn held between the godets.

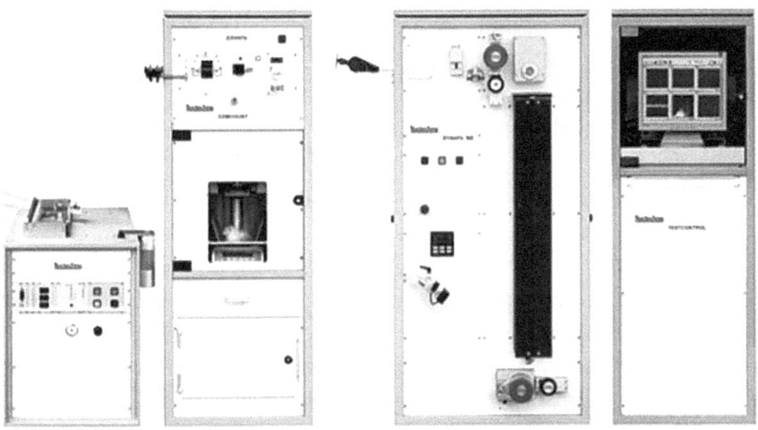

Fig. 6.8: DYNAFIL with COVAFIL, package changer and TESTCONTROL

The instrument incorporates a patented high-temperature heater that operates between the room temperature and 500°C. In conjunction with godet yarn feed mechanisms and electronically regulated servo drives, it enables test speeds of up to 1000 m/min, which have until now been impossible to realise. Furthermore, it opens new application possibilities in the area of high-temperature resistant polymers. A further technical innovation is the high-resolution load cell for highly accurate measurement of even the smallest forces, for example, for crimp tests on the texturised yarns.

Apart from standard tests for routine quality control, the PC control technique also permits the use of complex test parameter linkages, which allow for substantially more comprehensive statements to be made concerning yarn quality and the sources of faults during production. Dynafil tests can often replace work-intensive knit-dye tests that are commonly used in texturising. A change in the test material, yarn count or testing method is PC-controlled and requires no manual adjustment to the test equipment.

The application areas of the Dynafil also include friction measurements, entanglement tests, yarn evenness tests and yarn count measurements using additional modules. At the end of the testing sequence, the yarn samples can be captured by an automatic sample collector and be forwarded for further testing, for example, for determining spin-finish or coning oil contents.

To achieve a high test output, an automatic changer for feeding the test packages is provided. The Textechno package changers model SM and SE with 2 or 20 positions, respectively, operate on the principle of splicing the running yarn. This method enables package change without stoppage of the test equipment during the package changeover.

6.2 Entanglement Testing

The presence of entanglements affects the performance of the texturised yarns. Traditionally, this is done by piercing the yarn with a needle and pulling it through the yarn till it encounters a compaction. The distance covered by the needle is then measured. This is a time-consuming test and has been replaced by automatic testers. Here, the piercing unit is automatically inserted into the yarn and the yarn advances at a speed adjustable as required with its tension being measured. As soon the piercing unit encounters a compaction, the tension in the yarn increases, and at pre-selected tension levels, the length between the two compaction points is determined and displayed. In addition, soft entanglement points are determined.

6.3 Wickability Tests

Natural fibres are preferred for fabrics that are meant for wearing next to skin because of their good moisture absorption properties. By texturising, the moisture absorbency can be added to a synthetic filament yarn which is normally a non-absorbent. Wicking property is introduced into a man-made fibre yarn by texturising. Wicking is the spontaneous flow of a liquid in a porous substrate driven by the capillary force. This flow in the porous medium caused by the capillary action is governed by the properties of the liquid such as surface-wetting forces and geometric configuration of the pore structure such as yarn construction, number of fibres in cross section, the randomness of internal structure, twist, fabric structure and loop formation especially in the air-jet-texturised yarn. So measuring wickability is an important test for the texturised yarns. Wickability of the air-jet-texturised yarn and their fabric can be measured by both horizontal and vertical wicking testing method.

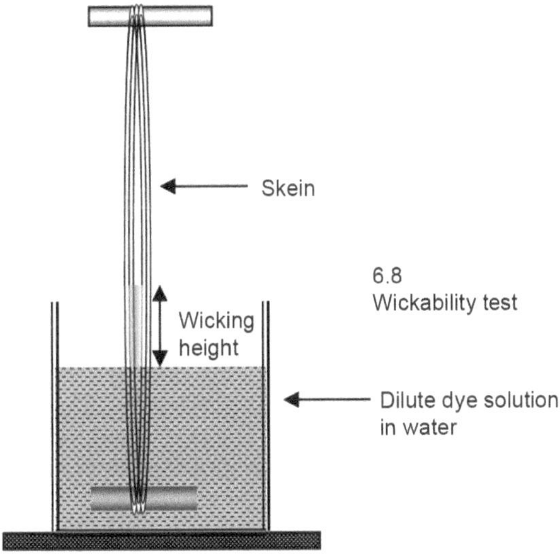

Fig. 6.9: Wickability Test

Wetting is a pre-requisite for wicking, which is of paramount practical importance in textile processes. 'Wettability' describes the interaction between the liquid and the substrate prior to wicking taking place. Most textile processes are time-limited, and the rate of wicking is therefore important. However, the wicking rate is not solely governed by interfacial tensions and the wettability of the fibres, but by other factors as well. The wicking rate depends on the capillary dimensions of the substrate and the viscosity of the liquid. The fibrous assemblies are usually considered to consist of a number of parallel capillaries. The advancement of the liquid front in a capillary can be visualised as occurring in small jumps. The advancing wetting line in a single capillary stretches the meniscus of the liquid until the elasticity of the meniscus and the inertia of flow are exceeded. The meniscus contracts, pulling more liquid into the capillary to restore the equilibrium state of the meniscus.

A lea is held tight with guide rods and the bottom of the lea is dipped in the dilute dye solution in water. The height of water penetration is noted after 15 min. The wickability is expressed as wicking height in centimetres (refer Figure 6.9).

6.4 Testing Interlaced Yarns

RIC interlace counter, developed by Schweizer consultants and promoted by PolySpinTex Inc., can be used for determining the number of nodes per metre, the average thickness of the yarn, the thickness of the not-interlaced yarn, the average node thickness, the yarn compaction factor, the node compaction factor, the tack effectiveness and the entanglement stability. Other standard values such as the average length of the tack and the CV% can also be evaluated by this instrument.

The appendixes explain various test methods and standards.

6.5 Evaluation of Physical Bulk of Air-Texturised Yarns

Bulk is an important characteristic of the air-texturised yarns. A number of machine yarn and process variations affect the yarn bulk. Because many of the fabric characteristics depend on the bulk, checking and controlling bulk is very important. The measurement of bulk of the texturised yarn cannot be done by actual measurement of the diameter using microscope or a thickness gauge. The normal practice is to evaluate the bulk by studying the package density of the final package wound with the texturised yarn.

$$\text{Physical Bulk (\%)} = \frac{\text{Density of parent yarn package}}{\text{Density of texturised yarn package}} \times 100$$

In case of using different types of yarns as a blend/mix (e.g. cotton, viscose, nylon, polyester), this formula is not applicable. Take the average of the number of type of yarn package.

$$\text{Physical Bulk (\%)} = \frac{\text{Average Density of parent yarn packages}}{\text{Density of texturised yarn package}} \times 100$$

Appendix

Appendix 1

ASTM D4031 – 07 (2012)

Standard Test Method for Bulk Properties of Texturised Yarns

Significance and Use

The values obtained by this test method should not be used to predict similar properties in the fabricated structures except in narrow well-defined comparisons, such as 16.7 tex (150 denier) polyester from the same feed yarn merge and texturised on the same machine type. Attempts to relate yarn performance to fabric performance might result in poor correlations unless other factors affecting bulk such as yarn shrinkage and fabric finishing are eliminated.

The elapsed time between processing and testing has a marked effect on the results of this test especially during the first 72 h. Therefore, specimens should only be compared if tested after the same elapsed time. This effect is caused by stress decay which is known to be minimal beyond the seventh day and after which time the sample remains relatively stable. Comparisons are preferably made after the seventh day.

In the case of yarns having a linear density near the upper limit of the skein size directed in Table 3, an error is introduced when rounding off to full revolutions. Therefore, the calculated values for crimp contraction and so on should only be compared with other samples of the yarn of the same linear density.

Option A used with crimp development Condition 1 (dry heat oven at 120°C (248 F)) and light loads of 0.04 mN/tex (0.5 mgf/den) and 0.44 mN/tex (5.0 mgf/den) is recommended for the texturised polyester yarns. All crimp parameters may be calculated.

Option B may also be used with crimp development Condition 1 (dry heat) for the texturised polyester yarns. Crimp contraction may be calculated. When used to duplicate or to utilise suitable mechanical yarn handling devices, alternate skein size and weights may be used as described in Chapter6.5.2.

Option C used with crimp development Condition 2 (water bath at 82°C (180 F)) and a light load of 0.13 mN/tex (1.5 mgf/den) is recommended for the texturised nylon yarns. For the texturised polyester yarns, Condition 3 (water bath at 97°C (206 F)) is recommended. Only bulk shrinkage is calculated.

This test method for the measurement of bulk properties is not recommended for acceptance testing of commercial shipments because of lack of precision data.

If there are differences or practical significance between the reported test results for two laboratories (or more), comparative tests should be performed to determine whether there is a statistical bias between them, using competent statistical assistance. As a minimum, test samples that are as homogeneous as possible drawn from the material from which the disparate test results were obtained and randomly assigned in equal numbers to each laboratory for testing. The test results from the two laboratories should be compared using a statistical test for unpaired data, at a probability level chosen prior to the testing series. If a bias is found, either its cause must be found and corrected, or future test results for that material must be adjusted in consideration of the known bias.

Table 3: Total Size (Linear Density) of Skein

Linear density of yarn	Linear density of skein
Options A, B, C: 1.7–44.4 tex (15–400 denier)	555.5 tex (5000 denier)
44.5–89.0 tex (401–800 denier)	833.5 tex (7500 denier)
Option B only (for mechanical device): 1.7–44.4 tex (15–400 denier)	250 tex (2250 denier)
Option C only: 1.7–44.4 tex (15–400 denier)[B]	

[B] 100 Revolutions, linear density of skein varies.

1. Scope

1.1 This test method covers the measurement of the change in the length of a tensioned skein of the texturised yarn due to the change in crimp characteristics brought about by exposure to wet or dry heat. The change in length, depending on procedure, is a measure of skein shrinkage, crimp contraction, bulk shrinkage or crimp recovery.

1.2 This test method applies to crimped, continuous multifilament yarns ranging from 1.7 to 88.9 tex (15–800 denier).

1.3 Three conditions are provided for crimp development mediums, and loading routines are provided to be used on the yarn skeins to allow determination of the yarn bulk by several different procedures.

1.4 The values stated in either SI units or inch-pound units are to be regarded as standard. Within the text, the inch-pound units are shown in parentheses. The values stated in each system are not exact equivalent; therefore, each system shall be used independently of the other. Combining values from the two systems may result in nonconformance with the specification.

1.5 This standard does not purport to address all the safety concerns, if any, associated with its use. It is the responsibility of the user of this standard to establish appropriate safety and health practices and determine the applicability of regulatory limitations prior to use.

2. Referenced Documents

ASTM Standards

D123 Terminology Relating to Textiles

D1059 Test Method for Yarn Number Based on Short-Length Specimens

D1776 Practice for Conditioning and Testing Textiles

D1907 Test Method for Linear Density of Yarn (Yarn Number) by the Skein Method

D2258 Practice for Sampling Yarn for Testing

D4849 Terminology Related to Yarns and Fibres

Appendix 2

ASTM D6774 – 02 (2010)

Standard Test Method for Crimp and Shrinkage Properties for Texturised Yarns Using a Dynamic Texturised Yarn Tester

Significance and Use

Test Method D6774 for determining the maximum total contraction, crimp and residual fibre shrinkage in the texturised filament yarns is suitable for acceptance testing of commercial shipments.

If there are differences in practical significance between the reported test results for two laboratories (or more), comparative tests should be performed to determine whether there is a statistical bias between them, using competent statistical assistance. As a minimum, use the samples for such comparative tests that are as homogeneous as possible, drawn from the same lot of material as the samples that resulted in disparate results during initial testing and randomly assigned in equal numbers to each laboratory. The test results from the laboratories involved should be compared using a statistical test for unpaired data at a probability level chosen prior to the testing series. If a bias is found, either its cause must be found and corrected, or future test results for that material must be adjusted in consideration of the known bias.

The properties and their variability as measured by this method relate to bulk appearance, stretch and recovery of the fabrics and dyeability of the yarns.

For some yarns, the elapsed time between processing and testing has a marked effect on the results, of this test, especially during the first 72 h. The effect is caused by stress decay which is known to be minimal beyond the seventh day and after which time the yarn remains relatively stable. Therefore, specimens should only be compared if tested after the same elapsed time. Samples can be tested at-line, thus having little to no elapsed time between processing and testing.

1. Scope

1.1 This test method covers the determination of crimp contraction, residual fibre shrinkage and their variability of all types of filament yarns (POY, FOY, flat yarns, texturised and bulked continuous filament (BCF) carpet yarns) using an automated tester.

Note 1: For another method of testing crimp in the texturised yarns, refer to Test Method D4031.

1.1.1 This method may also be used for the non-texturised yarns.

1.2 This test method is limited to crimped, multi-filament yarns ranging from 22.0 to 890 dtex (15–800 denier) and for BCF yarns from 890 to 4200 dtex (800–3800 denier).

1.3 The values stated in either SI or US customary units are to be regarded separately as standard. Within the text, the US customary units are in parentheses. The values stated in each system are not exact equivalents; therefore, each system shall be used independently of the other. Combining values from the two systems may result in nonconformance with the method.

1.4 This standard does not purport to address all the safety concerns, if any, associated with its use. It is the responsibility of the user of this standard to establish appropriate safety and health practices and determine the applicability of regulatory limitations prior to use.

2. Referenced Documents

ASTM Standards

D123 Terminology Relating to Textiles

D2258 Practice for Sampling Yarn for Testing

D4031 Test Method for Bulk Properties of Texturised Yarns

D4849 Terminology Related to Yarns and Fibres

Appendix 3

ASTM D2259 – 02 (2011)

Standard Test Method for Shrinkage of Yarns

Significance and Use

Test Method D2259 for testing the yarn for shrinkage in boiling water, saturated steam, dry heat or solvents is considered satisfactory for acceptance testing of commercial shipments of the yarn because the test method has been used extensively in the trade for that purpose.

If there are differences in practical significance between the reported test results for two laboratories (or more), comparative tests should be performed to determine whether there is a statistical bias between them, using competent statistical assistance. As a minimum, use the samples for such comparative tests that are as homogeneous as possible, drawn from the same lot of material as the samples that resulted in disparate results during initial testing and randomly assigned in equal numbers to each laboratory. The test results from the laboratories involved should be compared using a statistical test for unpaired data at a probability level chosen prior to the testing series. If a bias is found, either its cause must be found and corrected or future test results for that material must be adjusted in consideration of the known bias.

The results obtained by this test method can be used for the following purposes:

1. As an aid in predicting the dimensional stability of the fabrics to wet processing,
2. As an aid in predicting the dimensional stability of the fabrics during processing at elevated temperatures, and
3. As a control measure in the manufacture of some types of fibres.

The shrinkage medium to be used in the test depends on the requirements of the parties involved.

The procedure for shrinkage in boiling water is described in Section 12, for shrinkage in dry heat in Section 13, for shrinkage in saturated steam in Section 14 and for shrinkage in solvents in Section 15. A 30-min exposure time is prescribed for boiling water. In exposure to dry heat, saturated steam or solvents, the extent of the change in the length of the yarn is dependent on the fibre type and on the time and temperature of the exposure. Fibre types differ in their reaction to elevated temperature as well as the nature of the specific solvent, and prior fibre history can have a great influence on the heat and solvent shrinkage of a yarn. Therefore, the time and temperature conditions to be used to determine dry heat or saturated steam shrinkage must be agreed upon for the particular product involved. In addition, time and temperature conditions and solvent to be used must be agreed upon for solvent shrinkage determination for the particular product involved.

1. Scope

1.1 This test method is used to determine the shrinkage of the yarns in skein form when treated in boiling water, dry heat, saturated steam or solvents. This test method is applicable to the yarns made from any fibre or combination of fibres where the tex (denier or dtex) of the yarn is known or can be determined. This test method is not recommended for elastomeric yarns and those yarns that stretch more than 5% under the tension loadings prescribed, although it has been used for the latter.

Note 1: Procedures for determining yarn shrinkage and bulk properties of the texturised yarns are covered in Test Method D4031.

1.2 This test method shows the values in both SI and inch-pound units. 'SI' unit is the technically correct name for the system of metric units known as the International System of Units. 'Inch-pound' unit is the technically correct name for the customary units used in the United States. The values stated in either SI units or in other units shall be regarded as standard. The values expressed in each system may not be exact equivalents; therefore, each system must be used independently of the other without combining in any way.

1.3 This standard does not purport to address all the safety concerns, if any, associated with its use. It is the responsibility of the user of this standard to establish appropriate safety and health practices and determine the applicability of regulatory limitations prior to use.

2. Referenced Documents

ASTM Standards

D123 Terminology Relating to Textiles

D1059 Test Method for Yarn Number Based on Short-Length Specimens

D1776 Practice for Conditioning and Testing Textiles

D1907 Test Method for Linear Density of Yarn (Yarn Number) by the Skein Method

D2258 Practice for Sampling Yarn for Testing

D3888 Terminology for Yarn Spinning Systems

D3990 Terminology Relating to Fabric Defects

D4031 Test Method for Bulk Properties of Texturised Yarns

D4848 Terminology Related to Force, Deformation and Related Properties of Textiles

D4849 Terminology Related to Yarns and Fibres

E145 Specification for Gravity-Convection and Forced-Ventilation Ovens

Appendix 4

ASTM D2256/D2256M – 10e1

Standard Test Method for Tensile Properties of Yarns by the Single-Strand Method

Significance and Use

Acceptance testing – Option A1 of Test Method D2256 is considered satisfactory for acceptance testing of commercial shipments because the test method has been used extensively in the trade for acceptance testing. However, this statement is not applicable to knot and loop breaking force tests, tests on wet specimens, tests on oven-dried specimens or tests on specimens exposed to low or high temperatures and should be used with caution for acceptance testing because factual information on between-laboratory precision and bias is not available.

If there are differences in practical significance between the reported test results for two laboratories (or more), comparative tests should be performed to determine whether there is a statistical bias between them, using competent statistical assistance. As a minimum, use the samples for such comparative tests that are as homogeneous as possible, drawn from the same lot of material as the samples that resulted in disparate results during initial testing and randomly assigned in equal numbers to each laboratory. The test results from the laboratories involved should be compared using a statistical test for unpaired data at a probability level chosen prior to the testing series. If a bias is found, either its cause must be found and corrected or future test results for that material must be adjusted in consideration of the known bias.

Fundamental properties – The breaking tenacity, calculated from the breaking force and the linear density, and the elongation are the fundamental properties that are widely used to establish limitations on yarn processing or conversion and on their end-use applications. The initial modulus is a measure of the resistance of the yarn to extension at forces below the yield point. The chord modulus is used to estimate the resistance to imposed strain. The breaking toughness is a measure of the work necessary to break the yarn.

Comparison to skein testing – The single-strand method gives a more accurate measure of the breaking force present in the material than does the skein method and uses less material. The skein-breaking force is always lower than the sum of the breaking forces of the same number of ends broken individually.

Applicability – Most yarns can be tested by this test method. Some modification of clamping techniques may be necessary for a given yarn depending on its structure and composition. To prevent slippage in the clamps or damage as a result of being gripped in the clamps, special clamping adaptations may be necessary with high-modulus yarns made from fibres such as glass or extended chain polyolefin. Specimen clamping may be modified as required at the discretion of the individual laboratory providing a representative force–elongation curve is obtained. In any event, the procedure described in this test method for obtaining tensile properties must be maintained.

Breaking strength – The breaking strength of a yarn influences the breaking strength of the fabrics made from the yarn, although the breaking strength of a fabric also depends on its construction and may be affected by manufacturing operations.

Because breaking strength for any fibre type is approximately proportional to linear density, strands of different sizes can be compared by converting the observed breaking strength into breaking tenacity (centinewtons per tex, grams-force per tex or grams-force per denier).

Elongation – The elongation of a yarn has an influence on the manufacturing process and the products made. It provides an indication of the likely stretch behaviour of garment areas such as knees, elbows or other points of stress. It also provides design criteria for stretch behaviour of the yarns or cords used as reinforcement for items such as plastic products, hose and tires.

Force–elongation curve – Force–elongation curves permit the calculation of various values, not all of which are discussed in this test method, such as elongation at break, elongation at specified force, force at specified elongation, initial elastic modulus which is resistance to stretching, compliance which is the ability to yield under stress and is the reciprocal of the elastic modulus and area under the curve, a measure of toughness, which is proportional to the work done.

Note 3 – Force–elongation curves can be converted into stress–strain curves if the force is converted to unit stress, such as to centinewtons per tex, or pounds per square inch, or pascals, or gram-force per tex or gram-force per denier, and the elongation is based on the change per unit length.

Knot and loop breaking force – The reduction in the breaking force due to the presence of a knot or loop is considered a measure of the brittleness of the yarn. Elongation in knot or loop tests is not known to have any significance and is not usually reported.

Rate of operation – In general, the breaking force decreases slightly as time to break increases.

Operation of CRT, CRE and CRL tension testing machines at a constant time to break has been found to minimise differences in test results between the three types of tension testing machines. When tensile tests are performed at a fixed time to break, then reasonable agreement in the breaking force has generally been found to exist between CRT and CRE tension testing machines. Consistent results are also obtained between different manufacturers of CRL tension testing machines when they are operated at the same time to break. The agreement is not necessarily good, however, between CRE and CRT tension testing machines on one hand and CRL tension testing machines on the other even when they are all operated at the same time to break. The CRE-type tester is the preferred tension testing machine.

This test method specifies an average time to break of 20 ± 3 s as recommended by ISO TC 38 on Textiles, The International Standards Association test committee for standardising tests for fibres, yarns and fabrics. It also provides for alternate speeds, such as 300 ± 10 mm (12 ± 0.5 inch)/min when using a 250-mm (10-inch) gage length. The tolerance of ±3 s for the time to break is wide enough to permit convenient adjustment of the tension testing machine's rate of operation, and it is narrow enough to ensure good agreement between tests. The difference in the breaking force between tests at 17 and 23 s will usually not exceed 1.5% of the higher value.

In case a tension testing machine is not capable of being operated at 20-s time to break, alternative rates of operation are included in this test method. These alternative rates may be used only by agreement between the parties concerned or when required in an applicable material specification.

Tests on wet specimens – Tests on wet specimens are usually made only on the yarns which show a loss of the breaking force when wet or when exposed to high humidity, for example, the yarns made from animal fibres and man-made fibres based on regenerated and modified cellulose. Wet tests are made on flax yarns to detect adulteration by failure to show a gain in the breaking force. (This is not applicable to synthetic texturised yarns.)

Tests on oven-dried specimens and specimens at high temperatures – Tests on oven-dried specimens at standard or high temperatures are usually made only on the yarns that will be used at high temperatures or will be used under very dry conditions which will affect the observed breaking force, for example, on rayon yarns intended for use in tire cords and yarns for other industrial purposes. Note that the results obtained when testing oven-dried specimens at standard temperature will not necessarily agree with the results obtained when testing oven-dried yarns at high temperatures. (This is not applicable to synthetic texturised yarns.)

Tests on specimens at low temperatures – Tests on specimens exposed to low temperatures are usually made only on the yarns that will be used at low temperatures, for example, yarns used in outerwear designed for cold climates or outer-space situations. Low-temperature tests are made on the coated yarns used in the manufacture of the materials used in outdoor applications, such as screening fabrics.

1. Scope

1.1 This test method covers the determination of the tensile properties of monofilament, multifilament and spun yarns, which are single, plied or cabled with the exception of the yarns that stretch more than 5.0% when tension is increased from 0.05 to 1.0 cN/tex (0.5–1.0 gf/tex).

1.2 This test method covers the measurement of the breaking force and elongation of the yarns and includes directions for the calculation of breaking tenacity, initial modulus, chord modulus and breaking toughness.

1.2.1 Options are included for the testing of specimens in (A) straight, (B) knotted and (C) looped form.

1.2.2 Conditions of test are included for the testing of specimens that are (1) conditioned air, (2) wet, not immersed, (3) wet, immersed, (4) oven-dried, (5) exposed to elevated temperature or (6) exposed to low temperature.

Note 1 – Special methods for testing yarns made from specific fibres, namely, glass, flax, hemp, ramie and kraft paper, and for specific products, namely, tire cords and rope, have been published. Test Methods D885 and Specification D578. (This is not applicable to synthetic texturised yarns.)

Note 2 – For directions covering the determination of the breaking force of the yarn by the skein method, refer to Test Method D1578.

1.3 The values stated in either SI units or inch-pound units are to be regarded separately as standard. The values stated in each system may not be exact equivalents; therefore, each system shall be used independently of the other. Combining values from the two systems may result in non-conformance with the standard.

1.4 This standard does not purport to address all the safety concerns, if any, associated with its use. It is the responsibility of the user of this standard to establish appropriate safety and health practices and determine the applicability of regulatory limitations prior to use.

2. Referenced Documents

ASTM Standards

D76 Specification for Tensile Testing Machines for Textiles

D123 Terminology Relating to Textiles

D578 Specification for Glass Fibre Strands

D885 Test Methods for Tire Cords, Tire Cord Fabrics and Industrial Filament Yarns Made from Manufactured Organic-Base Fibres

D1578 Test Method for Breaking Strength of Yarn in Skein Form

D1776 Practice for Conditioning and Testing Textiles

D2258 Practice for Sampling Yarn for Testing

D2904 Practice for Interlaboratory Testing of a Textile Test Method that Produces Normally Distributed Data

D2906 Practice for Statements on Precision and Bias for Textiles

D4848 Terminology Related to Force, Deformation and Related Properties of Textiles

D4849 Terminology Related to Yarns and Fibres

E178 Practice for Dealing with Outlying Observations

Appendix 5

ASTM D6720 – 07 (2012)

Standard Test Method for Evaluation of Recoverable Stretch of Stretch Yarns (Skein Method)

Significance and Use

This test method is considered satisfactory for acceptance testing of commercial shipments because the current estimates of between-laboratory precision are acceptable and the method is used extensively in the trade for acceptance testing.

If there are differences in practical significance between the reported test results for two laboratories (or more), comparative tests should be performed to determine whether there is a statistical bias between them, using competent statistical assistance. As a minimum, use samples for such comparative tests that are as homogeneous as possible, drawn from the same lot of material as the samples that resulted in disparate results during initial testing, and randomly assigned in equal numbers to each laboratory. The test results from the laboratories involved should be compared using a statistical test for unpaired data, at a probability level chosen prior to the testing series. If bias is found, either its cause must be found and corrected or future test results for that material must be adjusted in consideration of the known bias.

This test differs from other crimp contraction test methods in that it measures the recoverable stretch during the unload cycle of the yarn.

1. Scope

1.1 This test method covers the determination of the recoverable stretch of the commercial texturised yarns, covered elastomeric yarns and other stretch yarns using skeins. This test method is particularly valuable for the yarns that develop additional crimp upon exposure to hot, wet conditions. The recoverable stretch is a relative measure of the recovery power the yarn can be expected to provide in a finished fabric.

1.2 This test method is applicable to continuous filament yarns and is suitable for the yarns that develop additional stretch potential upon exposure to heat.

1.3 This method is applicable to yarns 500 denier or finer.

1.4 The values stated in SI units are to be regarded as standard. The values given in parentheses are provided for information only.

1.5 This standard does not purport to address all the safety concerns, if any, associated with its use. It is the responsibility of the user of this standard to establish appropriate safety and health practices and determine the applicability of regulatory limitations prior to use.

2. Referenced Documents

ASTM Standards

D123 Terminology Relating to Textiles

D1776 Practice for Conditioning and Testing Textiles

D4849 Terminology Related to Yarns and Fibres

Appendix 6

ASTM D6612 – 00 (2011)

Standard Test Method for Yarn Number and Yarn Number Variability Using Automated Tester

Significance and Use

Test Method D6612 for yarn number and yarn number variability is satisfactory for acceptance of commercial shipments and is used in the trade.

If there are differences of practical significance between the reported test results for two or more laboratories, comparative tests should be performed by those laboratories to determine whether there is a statistical bias between them, using competent statistical assistance. As a minimum, samples used for each comparative test should be as homogeneous as possible, drawn from the same lot of material as the samples that results in disparate results during initial testing, and randomly assigned in equal numbers to each laboratory. Other fabrics with established tests values are used for this purpose. The test results from the laboratories involved should be compared using appropriate statistical analysis and a probability level chosen by the two parties before testing begins, at a probability level chosen prior to the testing series. If a bias is found, either its cause must be found and corrected or future test results adjusted in consideration of the known bias.

The average results from the two laboratories should be compared using appropriate statistical analysis and a probability level chosen by the two parties before the testing is begun. If a bias is found, either its cause must be found and corrected or the purchaser and the supplier must agree to interpret future test results with consideration to the known bias.

Test Method D6612 is also used for the quality control of the filament yarns.

Indices of variability:

Coefficient of variation – %CV is a standard statistical calculation and is the most common index of yarn unevenness. For most textile applications in the 80–330 dtex (70–300 denier) range, a 1.0–1.3%CV is adequate. %CV of the yarns coarser than 666 dtex (600 denier) is not routine and usually not meaningful. %CV is less discriminating than percent density spread (%DS).

Bad/good test – %BGT, which will normally be up to 20% greater than %DS value, emphasises the greatest spread in the entire length tested (%DS is an average). If the value is greater than 50% of the %DS, it suggests that there is a process that needs to be investigated.

Density spread – %DS is equivalent to the Uster % unevenness (Test Method D1425) and is an indication of short-term variability. The yarns with extreme values are more likely to cause trouble in the subsequent yarn processes, which makes this perhaps the most useful index. The minimum achievable and maximum tolerance spread for a yarn product will depend on the yarn manufacturing process and end use. A spread of 3%–4% generally is, for most textile applications, in the range of 160–550 dtex (150–500 denier). More critical applications, such as those using finer yarns, may require lower values.

Density frequency variability – DFV is an index of spacing variability, whereas the others are indices of magnitude or unevenness. Frequency variability can induce resonance in high-speed processing and is a common source of barre, dye streaks or patterned unevenness in the fabrics.

1. Scope

1.1 This test method covers the measurement of yarn number up to 4000 dtex (3600 denier) and related variability properties of the filament and spun yarns using an automated tester with capability for measuring mass variability characteristics.

1.2 Yarn number variability properties include %DS, %CV and density frequency variation.

Note 1 – For determination of yarn number by use of reel and balance, refer to Test Method D1907. For another method of measuring variability (unevenness) in the yarn, refer to Test Method D1425.

1.3 The values stated in either SI units or inch-pound units are to be regarded separately as standard. Within the text, the inch-pound units are in parentheses. The values stated in each system are not exact equivalents; therefore, each system shall be used independently of the other. Combining values from the two systems may result in inaccuracies of the results.

1.4 This standard does not purport to address all the safety concerns, if any, associated with its use. It is the responsibility of the user of this standard to establish appropriate safety and health practices and determine the applicability of regulatory limitations prior to use.

2. Referenced Documents

ASTM Standards

D123 Terminology Relating to Textiles

D1425 Test Method for Unevenness of Textile Strands Using Capacitance Testing Equipment

D1776 Practice for Conditioning and Testing Textiles

D1907 Test Method for Linear Density of Yarn (Yarn Number) by the Skein Method

D2258 Practice for Sampling Yarn for Testing

D4849 Terminology Related to Yarns and Fibres

Appendix 7

ASTM D4724 – 11

Standard Test Method for Entanglements in Untwisted Filament Yarns by Needle Insertion

Significance and Use

Option 1 of this test method for the determination of the degree of the untwisted filament yarn entanglement, as measured instrumentally, is used for acceptance testing of commercial shipments; however, caution is advised because information on between-laboratory precision is lacking. Comparative tests, may be advisable.

If there are differences in practical significance between the reported test results for two or more laboratories, comparative tests should be performed by those laboratories to determine whether there is a statistical bias between them, using competent statistical assistance. As a minimum, samples used for each comparative test should be as homogeneous as possible, drawn from the same lot of material as the samples that results in disparate results during initial testing, and randomly assigned in equal numbers to each laboratory. Other fabrics with established test values may be used for this purpose. The test results from the laboratories involved should be compared using appropriate statistical analysis and a probability level chosen by the two parties before testing begins, at a probability level chosen prior to the testing series. If a bias is found, either its cause must be found and corrected or future test results must be adjusted in consideration of the known bias.

Option 2 for this test method is intended for use when the supply of the yarn is limited.

The instrumental option of this test method is based on the total randomisation of the entanglements in the yarn; therefore, the distance measured between the point of insertion of a pin in the middle of the yarn and the point at which an entanglement is encountered, by movement of the yarn or the pin until it is stopped at a preset level of force, is representative of the distance between two entanglements at some location in the yarn.

Entanglements are used frequently instead of twist to ensure the integrity of the filament yarns. Such entanglements generally give somewhat less protection during weaving or knitting than twist, but with proper care will perform quite satisfactorily.

1. Scope

1.1 This test method covers two options for the measurement of entanglements in the untwisted filament yarns using needle insertion options for instrument (A) and manual (B) techniques.

1.2 The values stated in either SI or inch-pound units (in parentheses) are to be regarded separately as standard. The values stated in each system may not be exact equivalents; therefore, each system must be used independently of the other without combining values in any way.

1.3 This standard does not purport to address all the safety concerns, if any, associated with its use. It is the responsibility of the user of this standard to establish appropriate safety and health practices and determine the applicability of regulatory limitations prior to use.

2. Referenced Documents

ASTM Standards

D123 Terminology Relating to Textiles

D1776 Practice for Conditioning and Testing Textiles

D2258 Practice for Sampling Yarn for Testing

D4849 Terminology Related to Yarns and Fibre

Appendix 8

Standards under ISO/TC 38/SC 23 Secretariat for Texturised Yarns

ISO 2:1973 – Textiles – Designation of the direction of twist in yarns and related products

ISO 1139:1973 – Textiles – Designation of yarns

ISO 1144:1973 – Textiles – Universal system for designating linear density (Tex System)

ISO 2060:1994 – Textiles – Yarn from packages – Determination of linear density (mass per unit length) by the skein method

ISO 2061:2015 – Textiles – Determination of twist in yarns – Direct counting method

ISO 2062:2009 – Textiles – Yarns from packages – Determination of single-end breaking force and elongation at break using constant rate of extension (CRE) tester

ISO 2947:1973 – Textiles – Integrated conversion table for replacing traditional yarn numbers by rounded values in the Tex System

ISO 6741-1:1989 – Textiles – Fibres and yarns – Determination of commercial mass of consignments – Part 1: Mass determination and calculations

ISO 6741-2:1987 – Textiles – Fibres and yarns – Determination of commercial mass of consignments – Part 2: Methods for obtaining laboratory samples

ISO 6741-3:1987 – Textiles – Fibres and yarns – Determination of commercial mass of consignments – Part 3: Specimen cleaning procedures

ISO 6939:1988 – Textiles – Yarns from packages – Method of test for breaking strength of yarn by the skein method

ISO 8159:1987 – Textiles – Morphology of fibres and yarns – Vocabulary

ISO 8160:1987 – Textiles – Texturised filament yarns – Vocabulary

ISO 10132:1993 – Textiles – Texturised filament yarn – Definitions

ISO 18066:2015 – Textiles – Man-made filament yarns – Determination of shrinkage in boiling water

ISO 18067:2015 – Textiles – Synthetic filament yarns – Determination of shrinkage in dry hot air (after treatment)

Appendix 9

CTT – Yarn Appearance and Hairiness Tester

http://www.lawsonhemphill.com/LH-410-yarn-appearance-hairiness-tester.html

Introduction

The uniformity of the spun yarn has always been an important quality control parameter as it affects the commercial value of the yarn. Defects such as thick places, thin places, neps, hairiness and yarn appearance contribute to the overall yarn uniformity and altogether they define the character of the end product.

The CTT yarn analysis test offers very quick, precise yarn uniformity analysis based on the yarn diameter measurements. A CCD camera is used to measure the diameter values with 3.25 μm precision as the yarn is moving under constant tension at test speeds up to 360 m/min.

The unique camera design also provides the ability to test different types of synthetic yarns, including the high-performance yarns such as carbon, glass or hybrid yarns that cannot be tested with the capacitive-based systems.

Lawson-Hemphill Yarn Analysis Software, YAS, measures the yarn diameter and classifies the yarn defects such as thick places, thin places, neps and slubs by their diameter and length. The program provides the best simulated yarn appearance board in the market. The flexibility included in the YAS software allows the operator to change the board length and yarn fault descriptions without retesting the yarn. This allows the user to evaluate the same yarn at different loom widths to see whether any periodicity develops. Diameter data can also be exported to a fabric simulation software for further analysis.

Yarn types

Spun yarns (blend or 100%), ATYs

Instrument size

Depth	47 inches (119 cm)
Height	55 inches (140 cm)
Width	63 inches (160 cm)
Weight	760 lbs (345 kg)

Connections

Power: 110/220 VAC – 50/60 Hz

Ordering Information

Catalog number: LH-410 CTT-YAH

Appendix 10

Crimp Rigidity Tester

Crimp Rigidity of a texturised yarn gives a measure of its ability to recover from stretch. It is determined by measuring the percentage reduction in the length of a skein of the yarn immersed in water when a minor load replaces a major load.

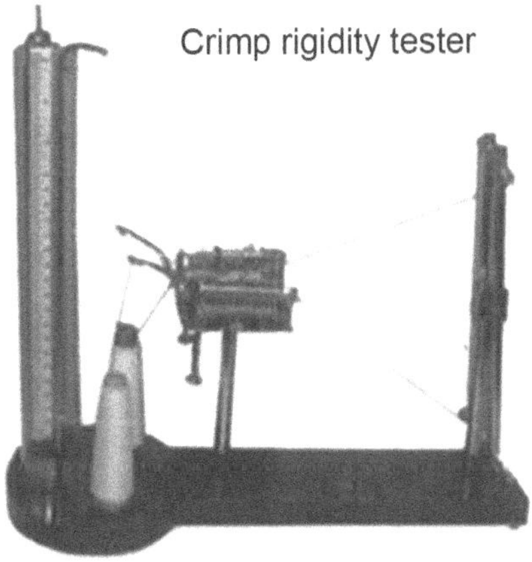

Crimp rigidity tester

For measuring crimp rigidity, a skein of known number of strands is prepared from the package of the yarn under test. This skein is then immersed in water filled in a glass jar with a load of 0.1 g/denier/strand suspended on it. The length of the skein is measured on a scale. The load is then reduced to 0.002 g/denier/strand by replacing the major load by the minor load and the reading on the scale observed. Crimp rigidity is calculated from the readings obtained under the major and minor loads.

References

1. H. V. Sreenivasamurthy, R. R. Dabhade, S. K. Sinha and S. P. Borkar. Proceedings of the seminar on 'Texturising Excellence' held at Daman TAI (M), 17 July 2004, pp. 9–21.
2. R. R. Dabhade (Guide H. V. Sreenivasamurthy). *Texturised Yarn Defects*. Undergraduate thesis submitted to University of Mumbai 1997–98.
3. S. Sinha (Guide H. V. Sreenivasamurthy). *Texturised Yarn Defects*. Postgraduate thesis submitted to University of Mumbai 1997–98.
4. Wikipedia – Free Encyclopedia.
5. Friction discs, http://www.polyspintex.com/education/friction-discs/
6. Tangled/Interlaced Yarns – Udo Schweizer's Texturising Pages, http://www.polyspintex.com/education/tangled-interlaced-yarn/
7. AYT – Air Jet Texturised Yarn – Udo Schweizer's Texturising Pages, http://www.polyspintex.com/education/ayt-air-jet-texturised-yarn/
8. Different Texturising Methods of Yarns – Textile learner blog spot, http://textilelearner.blogspot.com/2012/07/yarn-texturising-method-different.
9. Texturising, http://www.clothmen.com/dictionary.php?alpha=T&id=2775
10. C. Atkinson. *False Twist Texturised Yarns: Principles, Processing and Applications*. Woodhead Publishing Limited, 2012.
11. J. W. S. Hearle, L. Hollick and D. K. Wilson. Yarn Texturising Technology 2001, https://books.google.co.in/books?isbn=1855737655
12. V. B. Gupta and V. K. Kothari. Manufactured Technology – 1997, https://books.google.co.in/books?isbn=0412540304
13. Specifications for Texturised Yarn, http://www.saponamfg.com/texturisedspec.htm
14. D. Semnani and A. Gholami. *A Sharp Technique for Identification of Defective Points in False Twist Textured Yarns*. CSIR Publication, December 2009.
15. B. C. Goswami, J. C. Martindale and F. L. Scardino. *Textile Yarns: Technology, Structure and Application*. John Wiley, New York, 1977.
16. PolySpinTex, Inc. © 2014 Udo Schweitzer's Texturising Pages.
17. A. Grantham Pang. *Defect Detection in Texturised Material Using Gabor Filters*. IARL Publication, http://www.eee.hku.hk/
18. V. Usenko. *False Twisting Mechanisms – Processing of Manmade Fibers*. MIR Publishers, Moscow, 1979.
19. Symposium on Man-Made Fibres, IIT Delhi 1983.

20. P. Bajaj et al. *Cell Chem and Tech* 1978, 12, 387.
21. P. Bajaj et al. *Indian Journal of Textile Research* 1979, 4, 85.
22. P. Bajaj et al. *Journal of Textile Institute* 1980, 71, 284.
23. A. K. Sengupta et al. *Journal of Textile Institute* 1979, 70, 44.
24. A. K. Sengupta et al. *Journal of Textile Institute* 1979, 4, 91.
25. K. Jayachandran. *Man-Made Textiles*. Vol. 24 No. 9, September 1983.
26. K. Jayachandran. Texturising of cotton yarns. *Indian Textile Journal*, 1985.
27. B. Purushothama. *Effective Implementation of Quality Management Systems*. Woodhead Publishing, India, 2010.
28. B. Purushothama. *Work Quality Management*. Woodhead Publishing, India, 2013.
29. B. Purushothama. *Implementing ISO 9001;2015*. Woodhead Publishing, India, 2014.
30. Online textile education, www.aatcc.orgeventsonlineTextile-Fundamentals/demos/note books/Texturising pdf
31. Double Heater Stretch Yarn, http://www.polyspintex.com/education/double-heater-set-yarn/
32. M. Y. Gudiyawar Neha Hinge. Air jet texturising: a versatile process of texturising, www.2fashion
33. Texturised Yarns, http://encyclopedia2.thefreedictionary.com/Texturised+Yarns
34. H.R. Sheikh. Spinning review – texturised yarn production. *Pakistan Textile Journal*, 2011.
35. G. D. Wilkinson (Ed.). *Texturised Yarn Technology Vol I, Production and Properties and Processing*. Monsanto Publishing Co., 1967.
36. D. B. Ajagaonkar. *Texturised Yarns*. Part I, Vol. No. 2, Sept 1978, Part II, Vol. I, No. 5 December 1978, *Lohia Technical Services Bulletin*.
37. V. Subramanian et al. Development in texturising. *Indian Textile Journal*. No. 4, 1980.
38. C. J. Patel. Future of Texturising in India – Silk and Rayon Industries of India, No. 4, 1973.
39. W. Mey. Bulk, Stretch and Texture Process. Annual Conference of the Textile Institute, 1966.
40. M. R. Paranjape. Quality Control of Texturised Yarns – Man Made Textiles in India, March 1984.
41. A. T. Shukla. Cotton texturising. *Indian Textile Journal*. No. 4, 1985.
42. S. M. Ibrahim. Fundamentals of Core-Ply Texturised and Elastomeric Yarns. 53rd Annual Conference. The Textile Institute, 1968.
43. Barre in Fabrics, www.oocities.org/vijayakumar777/barre.htm
44. V. K. Kothari and N. B. Timble. Air texturising – effect of jet type and some process parameters on properties of air jet texturised yarns. *Indian Journal of Textile and Research*. Vol. 16, 1991.

45. G. R. Wrey. The properties of continuous air texturised yarns. *Journal of the Textile Institute*. 03/1969.

46. B. Purushothama. *Solutions to the Problems in Textile and Garment Industry*. Woodhead Publishing India, 2014.

47. Material Safety Data Sheet of Texturised Yarns, http://www.akra.com/media/hoja_seguridad_filamentos.pdf

48. T. R. Natarajan and E. Pio. *Quality Samurai – Designed Pathways for TQM Implementation*. Macmillan India Ltd, 1999.

49. S. Talawadekar. Visual Management through Five S: A Japanese Tool of Kaizen.

50. T. Lin, X. Wang and H. Wang. Self-crimping bicomponent nanofibres electrospun from polyacrylonitrile and elastomeric polyurethane.advanced materials. Vol. 17, Issue 22, Article first published online: 5 Oct 2005, http://onlinelibrary.wiley.com/

51. R. R. Hegde, A. Dahiya and M. G. Kamath (P. K. Jangala and R. Kotra). Bicomponent fibres, Updated April 2004, http://www.engr.utk.edu/mse/Textiles/Bicomponent%20 fibres.htm

52. Terms and Definitions – Dictionary.com

53. V. A. Usenko. Texturised yarns. *Pereabotka khimicheskikh volokon*. Moscow, 1975, pp. 255–396. *The Great Soviet Encyclopedia* (1979).

54. P. Bajaj, A. K. Sengupta and K. Sen. A study of the efficiency of catalysts used for texturising viscose by rapid curing. *Journal of the Textile Institute*, Vol. 71, Issue 6, 1980.

55. K. Sen. Texturising of cellulosic and polyester-cellulosic blends. Thesis submitted to the Indian Institute of Technology, Delhi, for the award of the Degree of Doctor of Philosophy. November 1980.

56. M. Yoshioka, N. Hagiwara and N. Shiraishi. Thermoplasticization of cellulose acetates by grafting of cyclic esters – Cellulose. September 1999, Vol. 6, Issue 3, pp 193–212.

57. A carpet with sheath/core BCF face yarns – US Patent WO 1998006562 A1.

58. German Hosiery Museum – Helanca, http://www.german-hosiery-museum.de/technik/garne/helanca/helanca.htm

59. False-twist texturising process with improved feed yarns and feed rates. US Patent – US 3816994 A

60. R. E. Whitfield, L. A. Miller and W. L. Wasley. Wool stabilization by interfacial polymerization, Part VIII. Location and nature of the polymer. *Textile Research Journal*. March 1965, Vol. 35, No. 3, 237–241.

61. TEXTURMAT ME+ Automatic crimp contraction and shrinkage tester for yarns, http://www.textechno.com/index.php/en/yarn-testing-products-76/texturmat-me

62. ASTM D4031 – 07 (2012) Standard Test Method for Bulk Properties of Texturised Yarns.

63. ASTM D6774 – 02 (2010) Standard Test Method for Crimp and Shrinkage Properties for Texturised Yarns Using a Dynamic Texturised Yarn Tester.

64. The TYT-EW and Texturised Yarn: New Quality Control Techniques, http://www.lawsonhemphill.com/pdf/lawson-hemphill-press-release-001.pdf

65. E. Unsal and A. Demir. Determination of the dye ability properties of PET texturised yarns by TYT equipment'. *Chemical Fibres International*, Vol. 51, 2001.

66. S. Ghosh and K. Carlan. Investigation on the use of on-line tension monitors to detect off-quality texturised yarns. *Chemical Fibres International*, Vol. 46, 1996.

67. B. Purushothama. Principles of Textile Testing and Statistics. The Textile Association (India), 2010.

68. M. Y. Gudiyawar Sultan H. Wicking Behavior of Air-Jet Blended Texturised Yarns and Their Fabrics By Fibre2Fashion.

69. Recent developments in the field of texturising technologies, Saturday, 6 June 2009, http://textilepoint.blogspot.com/2009/06/recent-developments-in-field-of.html

70. U. Mörschel, A. Paschen and W. Stein. Dynamic Testing on Drawn Texturised Yarns (DTY) with Dynafil ME. Textechno Herbert Stein GmbH & Co. KG.

71. B. Tex Laboratory Engineering. Crimp Rigidity Tester, https://www.textiletestingequipment.com/instrument-5/Yarn-Testing-Instruments/crimp-rigidity-tester.aspx

72. S. P. Singh. New hi-speed concept for making bi-shrinkage yarns. *Indian Textile Journal*, Jan 2007.

73. Crimp Rigidity Tester by Globe Enterprises, http://www.globetestingequipment.com/productlist.php?cid=11

74. ASTM D2259 – 02 (2011) Standard Test Method for Shrinkage of Yarns.

75. ASTM D2256/D2256M – 10e1 Standard Test Method for Tensile Properties of Yarns by the Single-Strand Method.

76. ASTM D6720 – 07 (2012) Standard Test Method for Evaluation of Recoverable Stretch of Stretch Yarns (Skein Method).

77. ASTM D6612 – 00 (2011) Standard Test Method for Yarn Number and Yarn Number Variability Using Automated Tester.

78. Standards and projects under the direct responsibility of ISO/TC 38/SC 23 Secretariat, http://www.iso.org/iso/iso_catalogue/catalogue_tc/catalogue_tc_browse.htm?commid=48334

79. MD Guidelines Occupational Information for Yarn-Texturing-Machine Operator I (Textile), http://74.63.141.36/yarn-texturising-machine-operator-i-textile

80. Md. Mominul Motin (Tusher). Optimization of productivity and maintenance in textile industries. *Bangladesh Textile Today.* February 2012, Vol. 05, Issue 02.

81. ASTM D4724 – 11 Standard Test Method for Entanglements in Untwisted Filament Yarns by Needle Insertion.

V. K. Kothari, A. Mukhyopadhay and R. C. D. Kaushik. Evaluation of physical bulk of air-texturised yarns. *Indian Journal of Fibre and Textile Research*. Vol. 27, March 2002.

List of Figures

About the Authors

H. V. Sreenivasamurthy is a retired Professor and Head of Textile Department, VJTI, Mumbai. He has served there for 32 years. He also served as the Chairman of Professionals Awards Committee, Textile Association (India), from 1985 to 2016, and Chairman, Textile Studies in Maharashtra State Technical Education Board. He has written a number of technical books and guided hundreds of students for their B. Tech and M. Tech projects. He is an approved guide for Ph.D. in Mumbai University. He has also published a number of articles in various national and international journals and conferences. He has the credit of introducing the G.M.T.A course and 3-year Diploma-level Associateship course in Textile Association, which is recognised as equivalent to a B. Tech degree offered by Technical Universities and Diploma courses offered by various state technical education boards. He is the fellow of the Textile Association (India) – FTA, and fellow of the Institution of Engineers (India) – FIE. He is also a Practicing Chartered Engineer and Textile Consultant. He is on the selection bodies and Board of Studies in UG and PG studies of various universities. He was a visiting faculty member of National Center of Quality Management (NCQM), Mumbai, and has successfully conducted a number of ISO-9000 awareness programmes in mills. From 2004 to 2009, he worked as an Advisor to Shri Vaishnav Institute of Technology & Science, Indore (M.P.), and consultant to SVK's – NMIMS – Centre for Textile Functions, Shirpur Campus (MH), since the year 2000. He is a recipient of Service Gold Medal and Service Memento of the Textile Association (India). He was presented the title of TAI Ratna on the platinum jubilee occasion of the Textile Association (India) on 9 April 2014. He is connected to various social and cultural organisations and has interests in literary activities, social services and yoga.

B. Purushothama is a postgraduate in textile technology, a fellow of Textile Association (India) and Institute of Engineers (India) and a Certified Quality Analyst with 43 years of industrial experience in textile industry. He is also a Lead Auditor for ISO 9001 and a Facilitator for Total Quality Management. He has written a number of technical and quality management books for textile industry and universities and is associated with a number of technical institutes and management institutes as a guest professor, syllabus committee member and examiner. He was a member of TX-30 committee of Bureau of Indian Standards for 14 years and a member of Spinning Advisory

Subcommittee of Bombay Textile Research Association (BTRA) for 8 years. He was one of the initial Quality Champions involved in implementation of JRD Quality Values and a Senior Examiner for Tata Business Excellency Module (TBEM) for 5 years. He has guided a number of B. Tech, M. Tech and MBA students for their project works. He is also involved in the activities of the Textile Association (India). He was presented the title of TAI Ratna on the platinum jubilee occasion of the Textile Association (India) on 9 April 2014. He is also active in the literary field in Kannada and has written a number of novels, poems, short stories and philosophical books in Kannada. He was awarded the prestigious 'Bharat Gaurav Award' on 11 March 2016 and "Bharat Ratna Dr. A. P. J. Abdul Kalam Excellence Award" on 14 Jan 2017 by India International Friendship Society in recognition of his lifetime achievements.

Index

D

E

F

G

H

I

J

K

L

M

N

O

P

Q

R

S

Y

Kolibri 360 GmbH
Ferdinand-Jühlke-Straße 7
99095 Erfurt, Deutschland
produktsicherheit@kolibri360.de